PAPUA NEW GUINEA

Mathematics

7A

Sue Gunningham
Pat Lilburn

OXFORD

OXFORD
UNIVERSITY PRESS

Level 8, 737 Bourke Street, Docklands, Victoria 3008, Australia.

Oxford University Press is a department of the University of Oxford. It furthers the University's objective of excellence in research, scholarship, and education by publishing worldwide. Oxford is a registered trademark of Oxford University Press in the UK and in certain other countries.

First published 2007
Reprinted 2008, 2009, 2010, 2013, 2014 (twice), 2023

ISBN 978 0 19 556248 4

Typeset by Damage Design
Illustrated by DiacriTech, Nives & Andy, and Uramina and Nelson Pty Ltd
Maps by Ophelia Leviny
Printed in Singapore by Markono Print Media Pte Ltd

Table of Contents

For Students

Dear Student,

We hope you enjoy working through the Oxford Mathematics books for Grade 7. The two books have been written to show how mathematics is useful in dealing with the everyday world. The lessons will help you to gain important life skills that will be useful now and in the future when you leave school to work for someone else or to set up your own small business.

The books comprise four topics:

Book A **Topic 1:** Look At Me

Topic 2: Our Country

Book B **Topic 3:** Work-wise

Topic 4: A Day in the Life

Each topic is broken into three *Learning Units* that will build on your previous knowledge and introduce you to new concepts. Each Learning Unit begins with pictures and open-ended questions to explore your knowledge of both the context and some of the mathematics to be covered.

Help Boxes have been located throughout the Learning Units to provide clear instructions to support your understanding of particular concepts. We encourage you to refer back to these Help Boxes as you progress through the book to refresh your memory and consolidate your understanding. A number of *Challenges* also appear in the Learning Units. It is hoped that you will attempt as many of these Challenges as possible to extend your thinking.

A Revision Unit, titled *Revision Challenges*, appears at the end of each topic after the three Learning Units. The Revision Unit provides an opportunity for you to demonstrate what you have learned and to undertake further investigations to deepen your understanding of the concepts covered in the Learning Units.

A *Glossary* appears at the back of each book. The words in the glossary relate specifically to the content of that particular book and the information provided should make it easier for you to understand the meaning of different mathematical terms.

An answer section has been included at the back of each book.

We hope you find the books an interesting and challenging part of your learning journey.

The Authors

Topic 1 Look At Me

Learning Unit 1: Food and Nutrition

Learning Unit 2: Healthy Living

Learning Unit 3: How Do I Compare?

Learning Unit 4: Revision Challenges

Topic 1 Look At Me

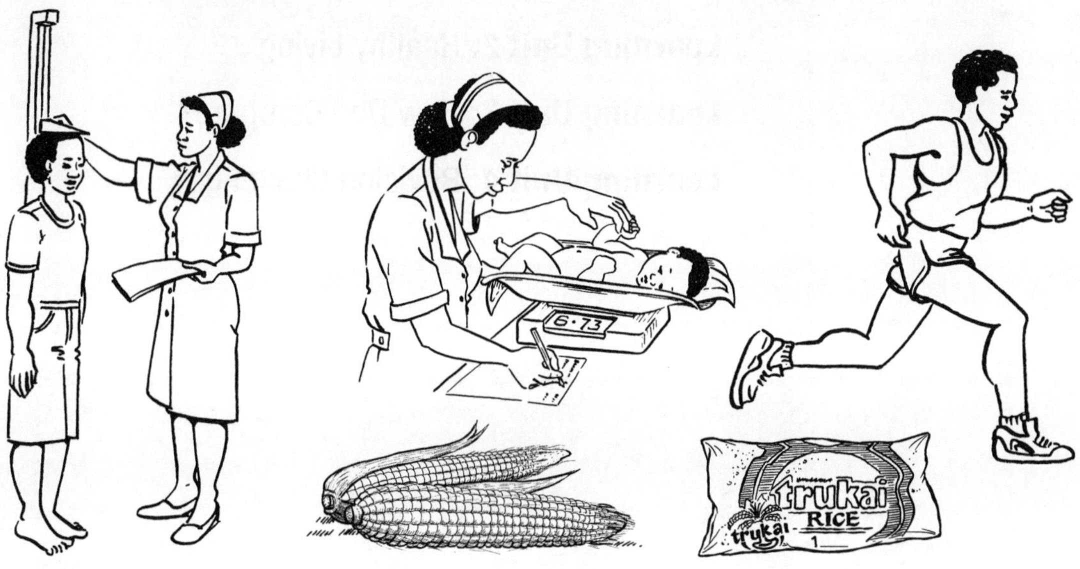

In this topic you will explore lots of things related to your own health and well-being.

The topic includes information about the importance of eating a balanced diet and doing enough physical exercise to maintain a healthy body.

A variety of fun tasks have been included that allow you to compare your own lifestyle and physical abilities against others.

Topic 1: *Look At Me* comprises the following Learning Units:

Overview

Each Learning Unit reflects an aspect of mathematics used in everyday life. A brief summary of the mathematics in each Learning Unit appears below.

In **Learning Unit 1**, *Food and Nutrition*, the work deals mainly with using the four operations and working with fractions, decimals and percentages. The unit also involves using estimation where appropriate to solve problems.

Learning Unit 2, *Healthy Living*, focuses on using and comparing data in tables and graphs, and classifying items in different ways using Venn diagrams.

The work in **Learning Unit 3**, *How Do I Compare?*, covers various measurements including weight, length, capacity and volume. Once again, estimation strategies are used where appropriate.

Learning Unit 4, *Revision Challenges*, uses a range of closed short-answer questions, open-ended questions and problems and short investigations to check your understanding of the topic overall.

Some of the materials that you may need to complete the activities in this topic are listed below.

You may need:

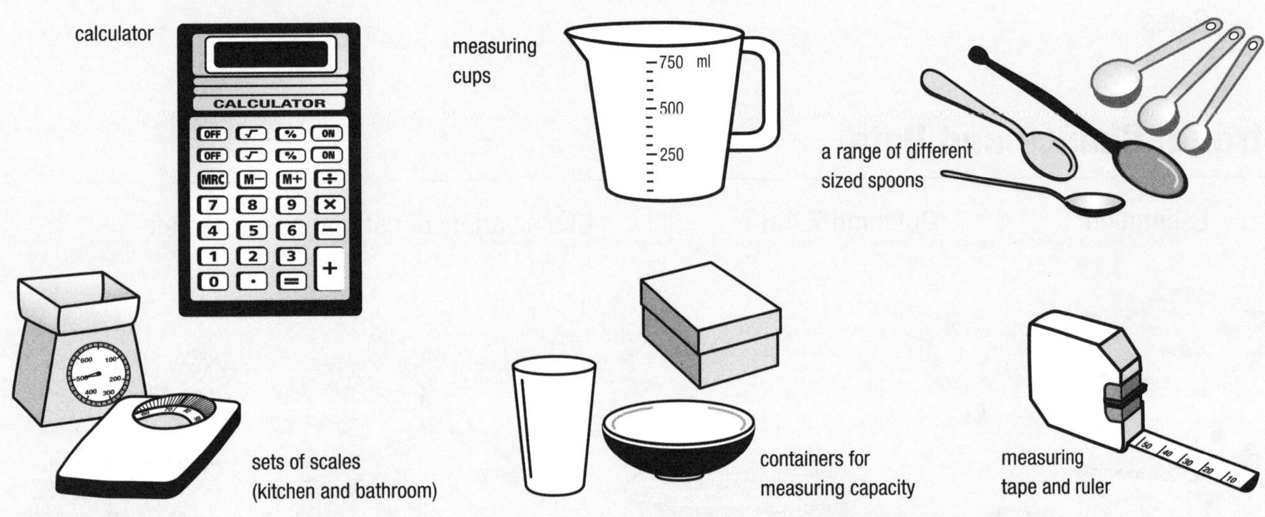

Learning Unit  Food and Nutrition

Strand: Number and Application

Fractions	Outcome 7.1.1	Solve problems requiring any of the four operations including mixed numbers
Decimals	Outcome 7.1.2	Use decimals in solving problems set in familiar contexts
Fractions and Decimals	Outcome 7.1.3	Convert between fractions, decimals and percentages
Decimals and Percentages	Outcome 7.1.4	Use percentages in a variety of real life situations
Ratios and Rates	Outcome 7.1.5	Convert between ratios and fractions

Strand: Chance and Data

Estimation	Outcome 7.4.6	Use a variety of estimation strategies

Lesson 1: Introduction	Looking at different foods and their purposes
Lesson 2: Food prices in Port Moresby	Reading bar graphs about average food prices Calculating possible percentage increases in food costs
Lesson 3: Kilojoules	Investigating the relationship between intake of kilojoules and exercise Calculating the decimal fraction of an hour needed to burn up kilojoules for a range of activities
Lesson 4: Reading food labels	Calculating differences between foods based on decimal information given on packages Calculating quantity of sugar per slice based on total sugar content of cakes
Lesson 5: Recipes and ratio	Using ratio to increase/decrease quantities in recipes Simplifying ratio Sharing profit based on ratio
Lesson 6: Strategies for estimation	Estimating by rounding to nearest $\frac{1}{10}$, 10, 100, 1000 Estimating when working with percentages Using front-end estimation
Lesson 7: Fractions of quantities	Understanding and representing fractions Calculating fractions of quantities Working with equivalent fractions
Lesson 8: Solving problems with fractions	Adding and subtracting fractions Multiplying fractions Dividing fractions
Lesson 9: Fraction, decimal and percentage conversions	Converting between fractions, decimals and percentages
Lesson 10: Solving food problems	Solving problems that involve fractions, decimals and percentages

Lesson 1 Introduction

During this unit you will revise some of the work you have already done with numbers including fractions, decimals and percentages. You will use your knowledge and skills in mathematics to make sense of information provided about food and nutrition. You will also learn about eating properly to stay healthy.

To stay healthy it is important to eat a 'balanced' diet. This means eating something from each of the main food groups every day.

The main food groups are:

- vegetables and fruit
- meat or beans
- bread or grains like rice and wheat
- dairy products like milk or cheese
- fats like oil or margarine

1 What foods do people in your family eat?

2 How many different types of food would you eat in a week?

3 List some of the foods that you eat for each of the five food groups.

4 Which food group do you eat the least?

5 Which food group do you like to eat most?

Each food group provides our body with different things. Some foods provide energy, some are body-building foods and some protect against disease. Energy foods like sago can fill stomachs quickly but may not be very nourishing. The following table lists some of Papua New Guinea's staple foods under these three categories.

Energy food	Body-building food	Protective food
coconut	eggs	sweet banana
sweet potato	fish	mango
taro	meat	apple
yam	tree nuts	pumpkin
sago		

6 What foods does your family grow?

7 What foods does your family buy?

8 Where do you think each of the five food groups fit under the headings 'energy, body-building and protective'?

Lesson 2 Food prices in Port Moresby

This bar graph shows the average price of certain foods in Port Moresby during 2006.

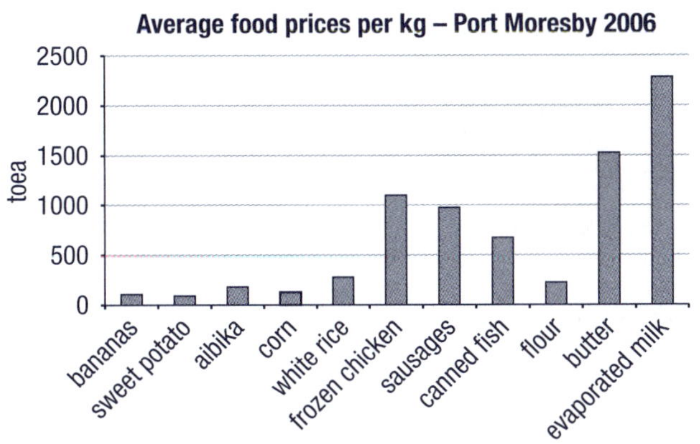

1 What was the average price of one kilogram of the following food items in Port Moresby in 2006?

a frozen chicken b aibika c flour d butter e bananas

2 Which of the five food groups seems to be the most expensive according to this graph? Why do you think that is?

3 Which of the five food groups seems to be the least expensive according to this graph? Why do you think that is?

4 a Name another food that your family eats regularly that is not shown on this graph.

b Estimate the average cost per kilogram of that food if it was bought in Port Moresby.

The cost of food in Port Moresby was about 5% higher in September 2006 compared to the same time in 2005. If this increase continues, food could become 5% more expensive each year. Usually the rise in the cost of food reflects the rise in the average wage.

Help Box

In Grade 6 you learned that percentage means out of 100.

To calculate 5% of a number, first find 10% (or $\frac{10}{100}$ which is the same as $\frac{1}{10}$) and then halve that amount (because 5% is half of 10%).

For example:
In 2006, onions cost 570t per kg. To find out what they might cost in 2007:

Find 10% of 570 = 57

So 5% of 570 = 28.5 (half of 57)
(Ignore the 0.5t for this estimation.)

In 2007, onions might cost 570t + 28t = 598t per kg.

5 How much did these foods cost in 2007 if they were 5% more than in 2006?
In 2006:

- **a** taro cost 250t per kg
- **b** coconuts cost 68t per kg
- **c** blade steak cost 1820t per kg
- **d** white bread cost 280t per loaf
- **e** tinned fish cost 350t per 500 g
- **f** biscuits cost 88t per 100 g

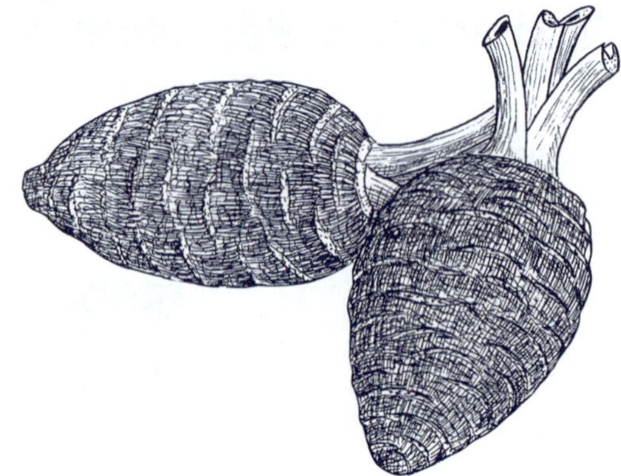

This graph shows the difference in prices that people pay for the same products at two different places in Papua New Guinea.

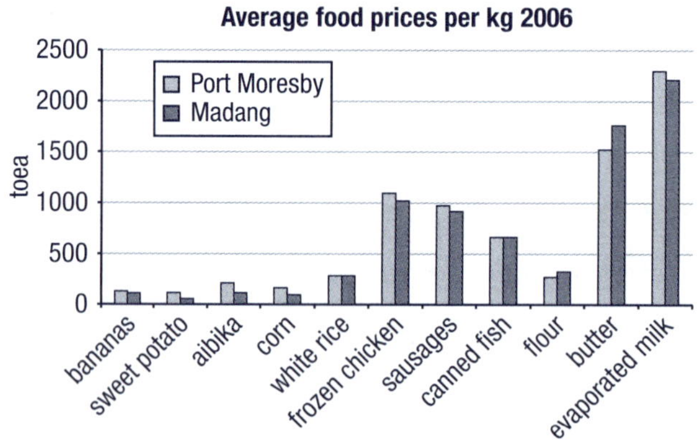

6 **a** Write about the differences in the average cost of sausages and frozen chicken between the two places.

b Write about the differences in the average cost of butter and flour between the two places.

7 Use the information in the graph to help you write a few sentences to explain some of the other differences between the average prices of different food products in the two places.

8 Why do you think there is a difference in price for the same product depending on where you buy it?

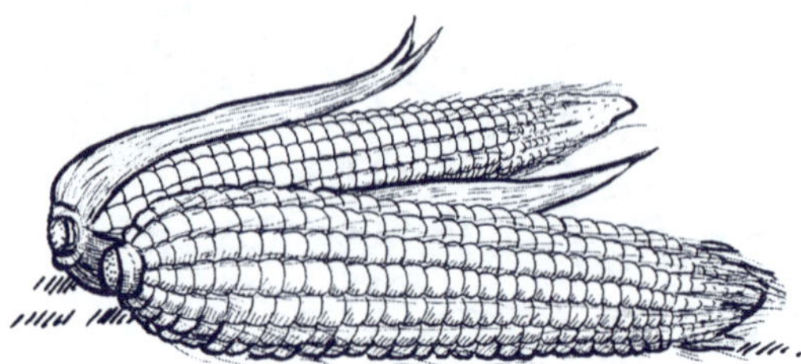

Lesson 3 Kilojoules

Different foods have different nutritional values. Kilojoules (kJ) is the measurement used to describe the 'energy' level contained in food. People who eat food high in kilojoules need to exercise to use up the 'energy' and prevent weight gain.

This table shows the kilojoules contained in some different foods.

Food	1 tomato	1 small carrot	1 boiled egg	1 slice pineapple	1 sausage	1 slice bread	110 g sardines	1 serve potato chips
kJ	85	90	335	120	700	360	447	1150

1 Order the foods from least to most kilojoules.

2 Describe two different combinations of foods from those listed that would contain a total of about 1000 kJ.

3 Think about the last meal you ate. Use your knowledge of the food groups and the information in the chart above to help you estimate how many kilojoules it might have contained.

Help Box

The table after this box shows that swimming uses 1150 kJ per hour. To work out the amount of time a person needs to swim to use the 360 kJ in a slice of bread, we need to divide the kJ in a slice of bread by 1150.

Method 1 (written):

Step 1: 115Ø ⟌ 36Ø

Step 2: $115\overline{)36}$ = 0

Step 3: $115\overline{)360}$ = 0.3

Step 4:
$$\begin{array}{r} 0.3 \\ 115\overline{)360} \\ -345 \\ \hline \end{array}$$

Step 5:
$$\begin{array}{r} 0.3 \\ 115\overline{)360} \\ -345 \\ \hline 15 \end{array}$$

Step 6:
$$\begin{array}{r} 0.30 \text{ of an hour} \\ 115\overline{)360} \\ -345 \\ \hline 15 \end{array}$$

0.30 of 1 hour = 0.3 × 60 = 18 minutes

Method 2 (calculator): Using a calculator, you can calculate the exact amount of time needed.

For example: 360 ÷ 1150 = 0.31

This means that the 360 kJ would be used in 0.31 of one hour of swimming.

Kilojoules used each hour for different activities			
Activity	**Number of kilojoules used in 1 hour**	**Activity**	**Number of kilojoules used in 1 hour**
sleeping	275	walking	900
sitting	380	swimming	1150
school work	550	gardening	1400
light activity	630	playing volleyball	1750
cleaning	800	running	1880

4 Use either of the methods shown in the Help Box to find out how long a person would have to swim to use up the kJ in:

- a one tomato
- b one boiled egg
- c 110 g of sardines on a slice of bread
- d four slices of pineapple
- e one carrot

5 Use either of the methods shown in the Help Box to find out how long a person would have to run to use up the kJ in:

- a one sausage
- b one carrot
- c one serve of potato chips
- d two slices of pineapple
- e a boiled egg, a sausage and a serve of chips

6 Choose a food from the table that you like to eat and work out how long it would take you to use up the kJ in that food if your only exercise was doing school work.

Challenge

Work out how many kJ a person would use if they:

- swam for 0.75 hour
- ran for 1.75 hours
- walked for 2.6 hours
- played volleyball for 1.3 hours
- sat for 3.2 hours
- cleaned for 1.9 hours

Lesson 4 Reading food labels

Food labels provide important information about the nutritional value of each product.

Bread 650 g NET
Serving size: 56.5 g (1 slice)
Energy 282.5 kJ
Protein 2.3 g
Fat – Total 0.4 g
Carbohydrates – Total 12.75 g
– Sugar 0.6 g
Sodium 147 mg

Serving size: 4 dry biscuits
Energy 209 kJ
Protein 1.2 g
Fat – Total 1.1 g
Carbohydrates – Total 8.8 g
– Sugar 0.2 g
Sodium 66 mg

1 Copy and complete this chart to show the differences between a serving size of the bread and a serving size of the dry biscuits.

	Kilojoules per serve	Protein per serve	Fat per serve	Sugar per serve	Sodium per serve
Bread (1 slice)					
Biscuits (4)					
Difference					

2 During the week, Kila eats two slices of bread for breakfast and uses two slices of bread to make a sandwich at lunchtime. Make a chart to show the total kJ, protein, fat, sugar and sodium that Kila eats each day just in the four slices of bread.

3 The nutrition label on a cake at the bakery says it contains 320 g of sugar. How much sugar will there be in each slice if the cake is cut into:

a 8 slices? b 10 slices? c 6 slices? d 12 slices?

4 Cakes containing different amounts of sugar have been cut into equal sizes. Copy and fill in the following table, showing how much sugar is in each slice of cake.

	kg of sugar per cake	slices in cake	kg of sugar per slice
a	0.280 kg	8	
b	0.360 kg	9	
c	0.720 kg	16	
d	1.250 kg	25	
e	1.5 kg	20	
f	2.4 kg	50	

Try to work out the answer without a calculator first.

$320 \text{ g} \div 6 = (300 \div 6) + (20 \div 6)$

$= (50) + (3.33)$

$= 53.33$ g or 0.053 kg

Then check your answer with a calculator.

Lesson 5 Recipes and ratio

Cooks need to be able to increase and decrease the ingredients in recipes to cater for different numbers of people.

This recipe makes 12 banana pikelets.

Ingredients

1 egg

1 mashed banana

600 mL milk

225 g self-raising flour

$\frac{1}{2}$ cup sugar

pinch of salt

Method

Beat the egg and sugar together.

Add the mashed banana.

Sift the flour and salt and add to the banana mixture.

Add the milk and mix to make a batter.

Heat a little butter in a pan.

Drop dessertspoons of batter into the pan.

Fry pikelets on one side then the other till golden brown.

1 How much flour would the cook need if he wanted to make the following number of pikelets?

a 6 pikelets b 18 pikelets

c 24 pikelets d 30 pikelets

2 How much milk would be needed to make the following number of pikelets?

a 6 pikelets b 18 pikelets

c 24 pikelets d 32 pikelets

3 How much flour would the cook need if he had bought the following number of eggs to make pikelets?

a 4 eggs b 6 eggs

c 12 eggs d 20 eggs

Help Box

We can use ratio when comparing two quantities.

For example:

Cordial is mixed using 1 part cordial concentrate to 3 parts water.

The ratio of cordial concentrate to water is 1 to 3. This can be written as 1:3.

Ratios can also be written as fractions. The fraction showing the ratio of cordial concentrate to water is $\frac{1}{3}$.

The ratio of eggs to pikelets in the recipe is 1 to 12. This can be written as 1:12.

The fraction showing the ratio of eggs to pikelets is $\frac{1}{12}$.

4 Write each of the following as a ratio and as a fraction.

a the number of carrots to number of pineapples

b the number of bananas to number of onions

c the number of pineapples to number of carrots

d the number of onions to number of bananas

5 At the market there were 9 stalls selling vegetables, 3 stalls selling fish, 1 stall selling chickens and eggs and 2 stalls selling bread. Write the ratio for the following.

a the number of bread stalls to number of vegetable stalls

b the number of chicken and egg stalls to number of fish stalls

c the number of chicken and egg stalls to number of bread stalls

d the number of vegetable stalls to total number of stalls

Help Box

Ratios and fractions can sometimes be simplified.

For example:

If there were 3 apples and 12 oranges the ratio would be 3:12.

Both numbers can be divided by 3. (3 is a factor of both numbers.)

The fraction $\frac{3}{12}$ can be simplified to $\frac{1}{4}$ and the ratio 3:12 can be simplified to 1:4.

6 Write each of these in their simplest form.

a 2:6 b 3:9

c 2:12 d 5:15

e 7:21 f 10:40

g 14:42 h 15:50

i 27:63 j 28:400

7 Write each of these in their simplest form.

a 5 g to 30 g b 10 kg to 30 kg

c 30 g to 750 g d 500 mL to 750 mL

e 2 L to 6 L f 3 cups to 15 cups

8 Change the amounts in each pair to the same unit and then write as a ratio in its simplest form.

a 250 g to 1 kg b 250 mL to 2 L

c 1250 mL to 3 L d 750 g to 2.5 kg

e $\frac{1}{2}$ kg to 800 g f 2 kg to 50 g

g 75 kg to 1 tonne h 1250 mL to $5\frac{1}{2}$ L

9 Leti made one dozen banana scones from 4 bananas, 3 cups of grated coconut and 6 cups of sago. How many bananas would Leti use if she made the following number of scones.

a 24 scones b 18 scones c 6 scones

10 Write the following ratios in Leti's scone recipe.

a number of bananas to number of cups of sago

b cups of coconut to cups of sago

11 A rectangular baking tin is sold in four different sizes using the same ratio.

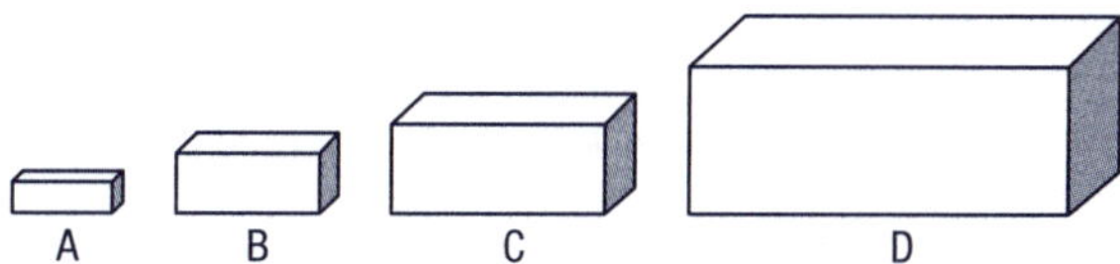

Copy and complete the table below to show the measurements for each sized tin.

Baking tin	Width	Length	Ratio
A	8 cm	12 cm	2:3
B	12 cm		2:3
C		24 cm	2:3
D	20 cm		2:3

Joseph and Dorcas had a stall at the food market. Joseph worked for 3 hours each market day and Dorcas worked for 2 hours. They decided to split the profits using a 2:3 ratio. The first week, the stall made a profit of K150. The Help Box below explains the method they used to calculate the profit split.

Help Box

To share K150 in the ratio 2:3, first there must be 5 parts (2+3).

One share is $\frac{2}{5}$ of 150 $= \frac{2}{{}_{1}\not{5}} \times \frac{\not{150}^{\,30}}{1}$

$= 2 \times 30$

$=$ K60

The other share must be K90 (K150 − K60).

12 Work out answers to the following.

a share K200 in the ratio 1:4

b share K200 in the ratio 3:7

c share K500 in the ratio 2:3

d share K500 in the ratio 3:7

e share K90 in the ratio 2:4

f share K90 in the ratio 1:9

g share K100 in the ratio 3:7

h share K100 in the ratio 1:4

Here is an example of how to increase a recipe for 4 pikelets to serve more people.

The original recipe given here makes 4 large pikelets.	Increasing the recipe by using a 1:4 ratio will make 16 large pikelets.	To purchase these ingredients, round to the nearest appropriate figures.
220 g SR flour	880 g SR flour (0.88 kg SR flour)	1 kg SR flour
2 tbsp butter	8 tbsp butter	8 tbsp butter
1 egg	4 eggs	4 eggs
$\frac{3}{4}$ cup milk	$\frac{12}{4}$ cups milk (3 cups milk)	3 cups milk
$\frac{2}{3}$ cup sugar	$\frac{8}{3}$ cups sugar ($2\frac{2}{3}$ cups sugar)	3 cups sugar

13 What ratio would have to be used to increase the original recipe to make 24 pikelets?

14 Write a new list of ingredients using this ratio.

Lesson 6 Strategies for estimation

Sometimes we don't need to use exact numbers. Instead, we can estimate an answer by rounding numbers to the nearest 10, 100 or 1000. It is easier to compare some things by rounding the numbers involved to the same level of accuracy.

To round numbers, think about where the number is located on the number line.

To round 37 to the nearest 10 you need to decide if 37 is nearer to 30 or 40.

To round 238 to the nearest 100 you need to decide if 238 is closer to 200 or 300.

To round 2.47 to the nearest $\frac{1}{10}$ you need to decide if 2.47 is closer to 2.4 or 2.5.

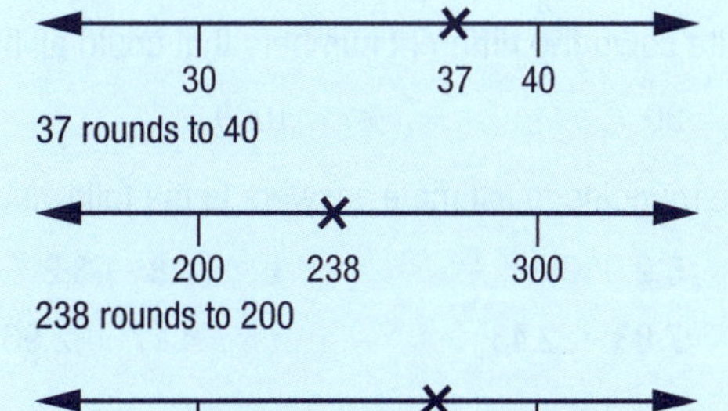

238 rounds to 200

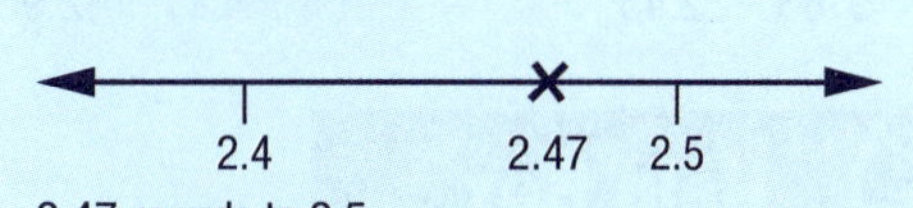

2.47 rounds to 2.5

1 Round each of these numbers to the nearest 10.

a 12 b 43 c 57 d 62
e 84 f 21 g 74 h 39

2 Round each of these numbers to the nearest 100.

a 431 b 297 c 854 d 662 e 848 f 259

3 Round each of these numbers to the nearest $\frac{1}{10}$.

a 2.47 b 59.62 c 47.08 d 1.36 e 92.86 f 31.59

We usually round numbers upwards if they are halfway between two numbers.

So, 15 rounded to the nearest 10 is 20, 450 rounded to the nearest 100 is 500 and 2.45 rounded to the nearest $\frac{1}{10}$ is 2.5.

4 Round these numbers to the nearest 10.

a 25 b 175 c 315 d 205 e 4065 f 2115

5 Round these numbers to the nearest 100.

a 150 b 350 c 850 d 1250 e 7850 f 4050

6 Round these numbers to the nearest $\frac{1}{10}$.

a 2.75 b 8.95 c 53.05 d 9.15 e 215.05 f 163.45

7 Kekeni said she had eaten 9743 kJ of food today. Round this amount to the nearest 100 kJ.

8 Kekeni said she had used up about 9000 kJ of energy during the day. How many kJ of energy might she actually have used?

9 The recipe said about 15 minutes preparation time. What might the preparation time actually be?

10 Write down five different weights that could all be rounded to:

a 2 kg b 45 g

11 Write down five different capacities that could all be rounded to:

a 650 mL b 1 L

12 Write down five different numbers that could all be rounded to:

a 30 b 1000

13 Use rounding to estimate answers to the following questions. Show the numbers that you used to work out your answers.

a $5.9 + 3.8$ b $7.8 - 3.2$ c 2.4×5.9 d $9.3 \div 3.2$

e $7.83 + 2.45$ f $4.17 + 2.96$ g 7.83×2.41 h $6.31 \div 1.75$

Help Box

Estimation can also be used when working with percentage. When using percentages to estimate, think about them in relation to known percentages such as 25%, 50% and 75%.

Examples:

To estimate 45% of 3000, think about 50% or $\frac{1}{2}$ of 3000. 50% of 3000 is 1500, so 45% must be a bit less.

To estimate 85% of 1600, think about 75% or $\frac{3}{4}$ of 1600. 75% of 1600 is 1200, so 85% must be a bit more.

To estimate 8% of 2950, think about 10% of 2950. 10% of 2950 is 295, so 8% must be a bit less.

14 Estimate answers to the following questions. Show the numbers that you used to work out your answers.

a 28% of 200 b 12% of 63

c 47% of 95 d 9% of 2468

Help Box

Another way to estimate answers is to use front-end estimation. This means only using the digits with the highest place value to calculate a rough estimate of the answer.

To estimate the answers to:

- 38 + 53, use the numbers 30 and 50 to make an estimate of 80.
- 337 − 83, use the numbers 300 and 80 to make an estimate of 220.

15 Use front-end estimation to estimate the answers to these additions and subtractions.

a 78 + 41 b 301 + 436 c 297 − 158 d 329 − 56

16 Use rounding to estimate the answers to the examples in question 15.

17 a Work out the exact answer for each example in question 15.

b Which method of estimation (front-end or rounding) was closest to the actual answers in each example?

18 Estimate answers to the following problems by using front-end estimation and rounding, and then work out the exact answer to determine which estimation method was closest to the actual answer.

a 243 + 549 b 276 + 589 c 397 + 207 d 513 + 239

e 334 ÷ 89 f 805 ÷ 185 g 659 × 82 h 1127 × 241

19 Estimate the answer to each problem by rounding to either 10 or 100.

a 378 + 921 + 246 b 76 + 432 + 1076 c 801 − 327 d 2983 − 345

e 624 ÷ 23 f 386 ÷ 169 g 723 × 18 h 1433 × 27

20 Elsie kept a list of the kJ she had eaten during the day. It looked like this:
260 + 115 + 650 + 85 + 335 + 1050 + 450.
Use front-end estimation to estimate the total kJ Elsie ate and then calculate the exact amount.

21 Elsie had also written down how much energy she thought she had used during the day.
It looked like this: 760 + 450 + 640 + 225 + 275.
Use rounding to estimate the total kJ Elsie used and then calculate the exact amount.

22 Look at the prices of food items on the right. Estimate the total cost of all the food using either rounding or front-end estimation and then calculate the exact cost.

23 Use front-end estimation to estimate answers to the following problems and then calculate the exact answers.

a 3.9 + 9.7 b 4.03 + 4.32

c 12.35 + 2.7 d 35.035 + 2.57

e 4.2 ÷ 2.1 f 12.42 ÷ 3.2

g 39.7 × 3.4 h 140.65 × 9.9

Lesson 7 Fractions of quantities

Help Box

What fraction of this grid has been coloured?

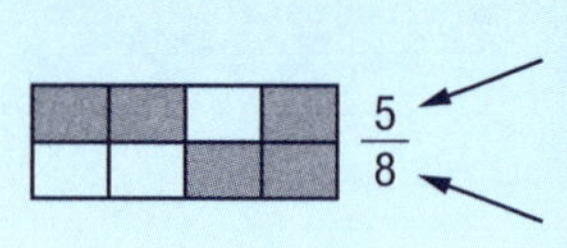

The **numerator** tells the number of parts being described.

The **denominator** tells the number of equal parts altogether.

What fraction of the beans has been split?

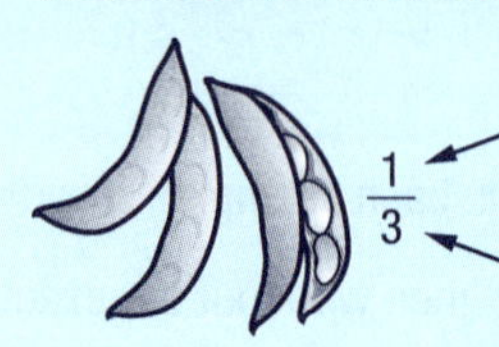

The **numerator** tells the number of parts being described.

The **denominator** tells the number of equal parts altogether.

When an item or a collection of items is divided into equal parts, each part is called a fraction of the item. The number on the bottom of the fraction (the denominator) tells how many parts the item or collection of items has been divided into. The number on the top (the numerator) tells how many parts are being described in the fraction.

1 Annette has a basket of vegetables. Five are green, three are yellow and two are red.

- a What fraction of the vegetables is green?
- b What fraction of the vegetables is yellow?
- c What fraction of the vegetables is red?

Help Box

Fractions can also be represented on a number line. This number line is divided into 8 equal parts. Each section is $\frac{1}{8}$ of the whole line. The cross on the number line shows the fraction $\frac{5}{8}$.

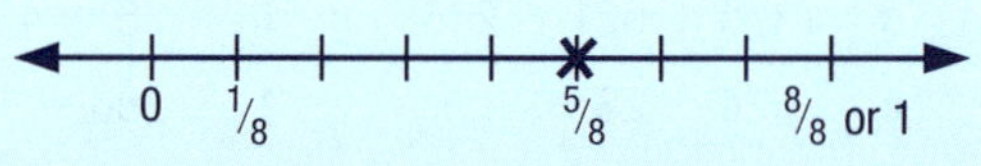

2 Draw number lines in your book to show the following fractions.

- a $\frac{3}{5}$
- b $\frac{2}{3}$
- c $1\frac{4}{7}$

3 Calculate.

- a $\frac{1}{8}$ of 16 berries
- b $\frac{3}{4}$ of 20 apples
- c $\frac{2}{5}$ of 15 bananas
- d $\frac{2}{3}$ of 30 oranges
- e $\frac{3}{7}$ of 28 fish
- f $\frac{5}{6}$ of 12 coconuts

4 The relationship between 100 mL and 1 L can be described by the statement '100 mL is $\frac{1}{10}$ of 1 L'. Write a fraction to describe the relationship in each of the following examples.

- a 100 g and 1 kg
- b 250 mL and 1 L
- c one toea and one kina
- d three buns and a dozen buns
- e 10 g and $\frac{1}{2}$ kg
- f two cans and one dozen cans

5 How many:

- a mL in $\frac{1}{2}$ L?
- b g in $\frac{3}{4}$ kg?
- c toea in $3\frac{1}{4}$ kina?
- d buns in $\frac{2}{3}$ of a dozen?
- e kg in $\frac{2}{5}$ tonne?
- f yams in $\frac{5}{6}$ of 4 dozen?

6 a How many 1 L containers of milk would be needed to fill ten 200 mL bottles?

b What fraction is one 200 mL bottle of 1 L?

7 a How many 250 mL bottles of cordial can be filled from a 3 L jug?

b What fraction of the jug is each bottle?

8 a How many 200 mL bottles of water can be filled from a 10 L container?

b What fraction of the container is each bottle?

Some fractions represent the same amount but have different names. They are called **equivalent fractions**. For example: $\frac{1}{2}$ and $\frac{2}{4}$ and $\frac{4}{8}$ are equivalent fractions.

This diagram shows that the three fractions describe the same amount of an object.

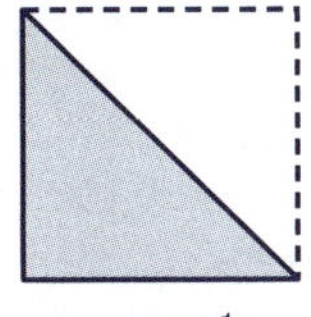

one half ($\frac{1}{2}$)

two quarters ($\frac{2}{4}$)

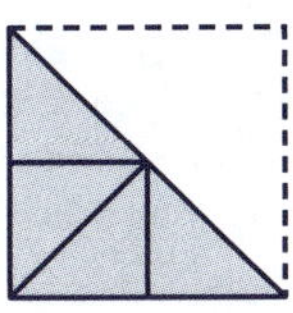

four eighths ($\frac{4}{8}$)

On a number line this can be shown in the following way:

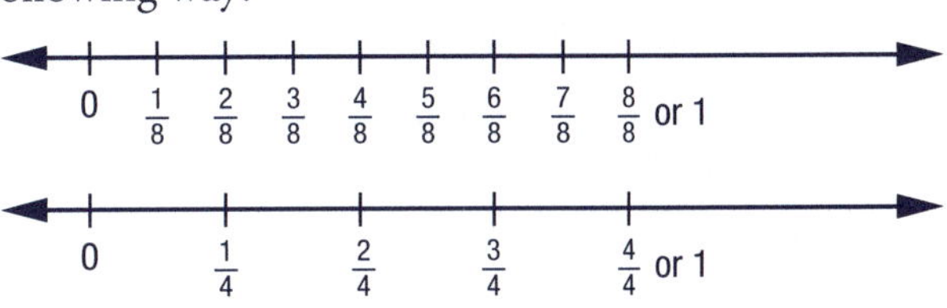

The diagrams show that $\frac{4}{8}$ and $\frac{2}{4}$ are different ways of describing $\frac{1}{2}$.

Help Box

We can create equivalent fractions by dividing the same number into both the numerator and the denominator.

$$\frac{5 \div 5}{10 \div 5} = \frac{1}{2},$$

so $\frac{5}{10}$ is the same as $\frac{1}{2}$.

Equivalent fractions can also be made by multiplying the numerator and the denominator by the same number.

$$\frac{3 \times 2}{8 \times 2} = \frac{6}{16},$$

so $\frac{3}{8}$ is the same as $\frac{6}{16}$.

9 Copy this table into your book and complete it by writing two equivalent fractions for each fraction listed.

Fraction	$\frac{2}{4}$			$\frac{12}{20}$			$\frac{10}{30}$		$\frac{3}{5}$	
Equivalent fraction	$\frac{1}{2}$	$\frac{3}{4}$	$\frac{2}{3}$		$\frac{7}{10}$	$\frac{4}{7}$		$\frac{20}{50}$		$\frac{15}{20}$
Equivalent fraction										

10 Write the following fractions in their simplest form.

a $\frac{3}{6}$ b $\frac{10}{30}$ c $\frac{12}{48}$ d $\frac{10}{100}$ e $\frac{15}{100}$

f $\frac{25}{100}$ g $\frac{18}{100}$ h $\frac{84}{100}$ i $\frac{6}{1000}$ j $\frac{85}{1000}$

11 Use what you know about equivalent fractions to replace the △ with >, < or =.

a $\frac{3}{10} \triangle \frac{7}{100}$ b $\frac{3}{7} \triangle \frac{15}{21}$ c $\frac{2}{5} \triangle \frac{15}{20}$ d $\frac{71}{100} \triangle \frac{9}{20}$

e $\frac{13}{15} \triangle \frac{20}{60}$ f $\frac{2}{13} \triangle \frac{15}{39}$ g $\frac{3}{4} \triangle \frac{15}{20}$ h $\frac{4}{5} \triangle \frac{9}{20}$

Lesson 8 Solving problems with fractions

Sometimes we need to add and subtract mixed numbers. This is easy if the fractions have the same denominator.

For example: $3\frac{1}{4}$ kg of sago + $1\frac{3}{4}$ kg of sago = 5 kg of sago.

However, to add and subtract fractions with unlike denominators, we need to first convert one or both fractions to an equivalent fraction so that they share the same denominator.

Help Box

Method 1: Addition of mixed numbers

Example: $2\frac{3}{4} + 1\frac{11}{12}$ can be solved by first adding the whole numbers $2 + 1 = 3$.

The problem now becomes $3 + (\frac{3}{4} + \frac{11}{12})$.

$\frac{3}{4}$ is equivalent to $\frac{9}{12}$, so $3 + \frac{9}{12} + \frac{11}{12} = 3\frac{20}{12}$.

$3\frac{20}{12}$ can be simplified to $4\frac{2}{3}$.

1 Use what you have learned to help you solve these problems.

a $1\frac{2}{3}$ kg fish + $1\frac{1}{6}$ kg fish

b $1\frac{3}{4}$ L cordial + $2\frac{1}{2}$ L cordial

c $\frac{3}{4}$ kg sago + $1\frac{3}{8}$ kg sago

d $2\frac{3}{5}$ L cooking oil + $\frac{7}{10}$ L cooking oil

2 Add these mixed numbers.

a $1\frac{1}{2} + 2\frac{7}{8}$

b $5\frac{1}{5} + 2\frac{9}{10}$

c $3\frac{11}{12} + 2\frac{3}{4}$

d $4\frac{7}{8} + 1\frac{1}{2}$

Help Box

Method 2: Addition of mixed numbers

Sometimes it is more difficult to find a common denominator for both fractions when adding mixed numbers.

Example: $2\frac{3}{4} + 3\frac{2}{3}$

First add the whole numbers: $2 + 3 = 5$

The problem now becomes $5 + \frac{3}{4} + \frac{2}{3}$.

Think of a denominator that could be used to convert both $\frac{3}{4}$ and $\frac{2}{3}$ to equivalent fractions. The lowest common denominator in this case would be 12.

$\frac{3}{4}$ can be converted to $\frac{9}{12}$ and $\frac{2}{3}$ can be converted to $\frac{8}{12}$.

The problem now becomes $5 + \frac{9}{12} + \frac{8}{12} = 5\frac{17}{12}$, which can be simplified to $6\frac{5}{12}$.

3 Add these mixed numbers by converting the fractions to equivalent fractions with the same denominator.

a $3\frac{2}{3} + 2\frac{7}{8}$ b $1\frac{1}{5} + 2\frac{3}{4}$ c $3\frac{1}{2} + 2\frac{2}{5}$ d $4\frac{7}{8} + 1\frac{3}{10}$ e $1\frac{1}{2} + 2\frac{2}{3}$

f $1\frac{3}{8} + 1\frac{7}{10}$ g $4\frac{4}{5} + 1\frac{1}{4}$ h $2\frac{1}{2} + 1\frac{1}{3}$ i $2\frac{5}{8} + 2\frac{1}{5}$ j $4\frac{7}{10} + \frac{7}{8}$

Help Box

Subtraction of mixed numbers

To subtract mixed numbers, first convert both numbers to improper fractions with the same denominator.

Example: $2\frac{3}{4} - 1\frac{11}{12}$ can be solved by first converting both numbers to improper fractions.

The problem now becomes $\frac{11}{4} - \frac{23}{12}$.

Convert $\frac{11}{4}$ to an equivalent fraction with a denominator of 12: $\frac{33}{12}$

The problem now becomes $\frac{33}{12} - \frac{23}{12} = \frac{10}{12}$, which can be simplified to $\frac{5}{6}$.

4 Use what you have learned to help you solve these problems.

a $2\frac{1}{3}$ kg salt − $1\frac{1}{4}$ kg salt
b $2\frac{3}{4}$ kg hamburger mince − $1\frac{1}{3}$ kg hamburger mince
c $1\frac{1}{2}$ L cola − $\frac{3}{10}$ L cola
d $2\frac{1}{3}$ of a sack of rice − $\frac{2}{3}$ of a sack of rice

5 Subtract these mixed numbers by converting to improper fractions with the same denominator.

a $2\frac{3}{4} - 2\frac{1}{8}$ b $5\frac{3}{4} - 2\frac{1}{8}$ c $4\frac{5}{12} - 2\frac{1}{4}$ d $4\frac{7}{10} - 1\frac{2}{5}$

e $2\frac{3}{5} - 1\frac{1}{3}$ f $3\frac{1}{2} - 1\frac{7}{10}$ g $2\frac{7}{10} - 2\frac{2}{3}$ h $2\frac{7}{12} - 1\frac{1}{5}$

6 Mirou's bag contained $\frac{2}{3}$ kg of sago and $1\frac{3}{4}$ kg of rice. The empty bag weighed $\frac{1}{8}$ kg. What was the total weight of the bag and its contents?

7 Fegsley bought $1\frac{1}{2}$ kg of peanuts and then his friend gave him $\frac{3}{4}$ kg for free.

a How many kilograms of peanuts does Fegsley now have altogether?
b If there are 48 peanuts in each kilogram, how many did Fegsley buy and how many did his friend give him?

Help Box

Multiplication of fractions

Steven ate half a pineapple for lunch every day for one week.

This could be written mathematically as $\frac{1}{2} + \frac{1}{2} + \frac{1}{2} + \frac{1}{2} + \frac{1}{2} + \frac{1}{2} + \frac{1}{2} = 3\frac{1}{2}$.

An easier way to calculate this is to use multiplication, for example: $7 \times \frac{1}{2}$.

This can be written as $\frac{7 \times 1}{1 \times 2}$

When multiplying fractions, first multiply the numerators and then multiply the denominators: $\frac{7}{1} \times \frac{1}{2} = \frac{7}{2}$ or $3\frac{1}{2}$

If you are multiplying mixed numbers, it is important to change them to improper fractions first. For example: $2\frac{1}{3} \times 1\frac{1}{4}$ becomes $\frac{7}{3} \times \frac{5}{4}$.

8 Calculate and write the answer in its simplest form.

a $\frac{3}{10} \times 5$ b $\frac{3}{8} \times 7$ c $4 \times \frac{5}{6}$ d $3 \times \frac{7}{10}$ e $2 \times \frac{3}{4}$ f $\frac{2}{13} \times 5$
g $7 \times \frac{1}{6}$ h $\frac{2}{3} \times 10$ i $\frac{3}{5} \times 5$ j $8 \times \frac{3}{10}$ k $2 \times \frac{3}{7}$ l $\frac{25}{100} \times 3$

9 The recipe said that $\frac{3}{4}$ of a cup of corn was needed to make a large pie. Kila only wanted to make a small pie so she needed to halve the recipe. She needed to use one half of $\frac{3}{4}$ of a cup of corn.

a Draw a diagram to show how much corn Kila needed for the small pie.

b Calculate the fraction that Kila needs of one whole cup of corn.

10 Calculate and write the answer in its simplest form.

a $\frac{3}{10} \times \frac{1}{6}$ b $\frac{3}{8} \times \frac{3}{5}$ c $\frac{3}{10} \times \frac{5}{6}$ d $\frac{3}{4} \times \frac{7}{10}$ e $\frac{2}{13} \times \frac{3}{7}$
f $\frac{2}{3} \times \frac{25}{100}$ g $\frac{3}{10} \times \frac{2}{5}$ h $2\frac{1}{3} \times \frac{3}{4}$ i $1\frac{4}{5} \times 1\frac{9}{10}$ j $2\frac{1}{2} \times 3\frac{2}{5}$

Challenge

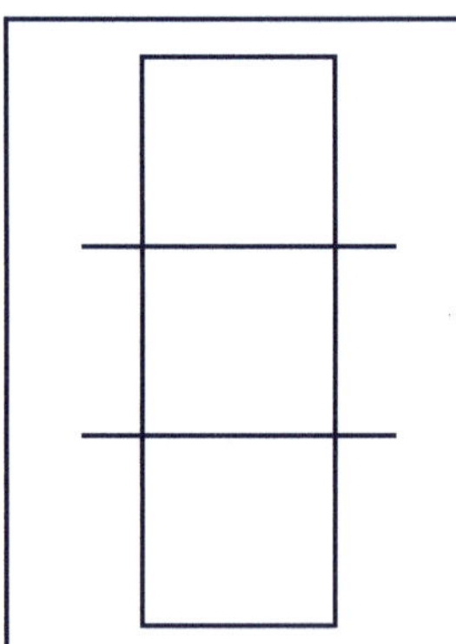

One third of a family's diet is made up of energy food.

Three quarters of that energy food are yams.

What fraction of the family's whole diet are yams?

Use the diagram to help solve this problem and then show how you worked out your answer mathematically.

Help Box

Division of fractions

To divide by fractions first invert the divisor, then replace the division sign with a multiplication sign and calculate the multiplication problem.

Example 1:

$\frac{3}{4} \div \frac{1}{2}$ can be solved by first inverting $\frac{1}{2}$ to become $\frac{2}{1}$.

Replace the $\div$ symbol with a $\times$ symbol: $\frac{3}{4} \times \frac{2}{1}$.

Calculate the answer $\frac{6}{4}$, which can be simplified to $1\frac{1}{2}$.

To divide mixed numbers, first convert both numbers to improper fractions then invert the divisor and multiply the fractions.

Example 2:

$2\frac{3}{4} \div 1\frac{1}{2}$

Convert the mixed numbers to improper fractions: $\frac{11}{4} \div \frac{3}{2}$

Invert the divisor and replace the $\div$ sign with a $\times$ sign: $\frac{11}{4} \times \frac{2}{3}$

Calculate the answer $\frac{22}{12}$, which can be simplified to $1\frac{5}{6}$.

11 Use what you have learned to help you solve these problems.

a Maria has $2\frac{3}{4}$ L of tomato pulp. How many $\frac{1}{2}$ L containers can she fill? ($2\frac{3}{4} \div \frac{1}{2}$)

b Peter has $2\frac{1}{3}$ kg of fish. How many $\frac{1}{4}$ kg serves will that make?

c Kali has $1\frac{2}{3}$ of a sack of rice. How many people will that feed if each person needs $\frac{1}{8}$ of a sack?

12 Calculate and write the answer in its simplest form.

a $\frac{9}{10} \div \frac{5}{6}$ b $\frac{3}{4} \div \frac{2}{5}$ c $\frac{7}{8} \div \frac{2}{3}$ d $\frac{3}{4} \div \frac{1}{8}$ e $\frac{2}{3} \div \frac{3}{4}$

f $\frac{4}{5} \div \frac{1}{8}$ g $\frac{2}{3} \div \frac{3}{10}$ h $\frac{3}{5} \div \frac{9}{10}$ i $\frac{1}{4} \div \frac{3}{5}$ j $\frac{3}{8} \div \frac{2}{5}$

13 Calculate and write the answer in its simplest form.

a $1\frac{2}{3} \div 2\frac{3}{10}$ b $2\frac{7}{10} \div 1\frac{5}{8}$ c $2\frac{9}{20} \div \frac{3}{4}$ d $1\frac{4}{5} \div \frac{9}{10}$ e $2\frac{11}{20} \div \frac{3}{7}$

Lesson 9 Fraction, decimal and percentage conversions

Help Box

Method 1: Converting fractions to decimals

In previous years you learned how to change fractions with denominators of 10 and 100 to decimals.

For example: $\frac{7}{10} = 0.7$; $\frac{33}{100} = 0.33$; $\frac{9}{100} = 0.09$

Fractions with denominators of 1000 can be changed in the same way.

For example: $\frac{378}{1000} = 0.378$; $\frac{96}{1000} = 0.096$; $\frac{7}{1000} = 0.007$

1 Change these fractions to decimals.

a $\frac{3}{10}$ b $\frac{47}{100}$ c $\frac{701}{1000}$

d $\frac{9}{100}$ e $\frac{37}{1000}$ f $\frac{263}{1000}$

2 Write these decimals as common fractions.

a 0.78 b 0.617 c 0.03

d 0.5 e 0.029 f 0.12

3 Write these decimals as common fractions in their simplest form.

a 1.3 b 3.06 c 7.004

d 3.2 e 4.05 f 10.75

g 2.37 h 4.125 i 7.26

j 2.075 k 5.004 l 22.255

These food containers have been marked to show their weight as a decimal fraction of a unit.

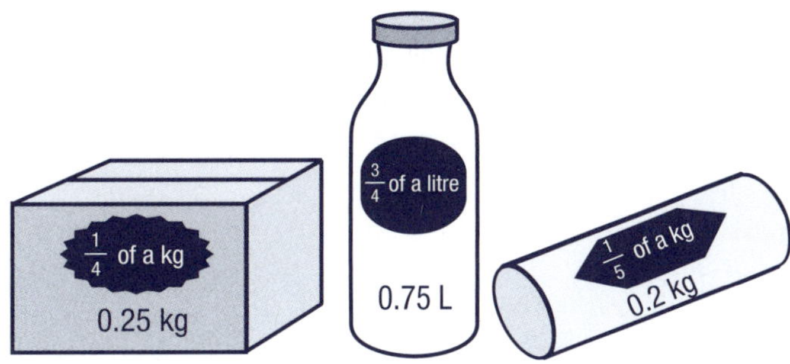

4 Write the weights of the food tins below as a decimal fraction of 1 kg.

a 

b 

c 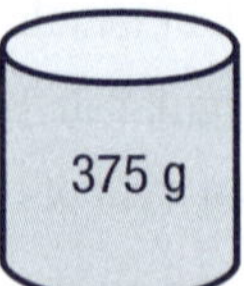

d 

e 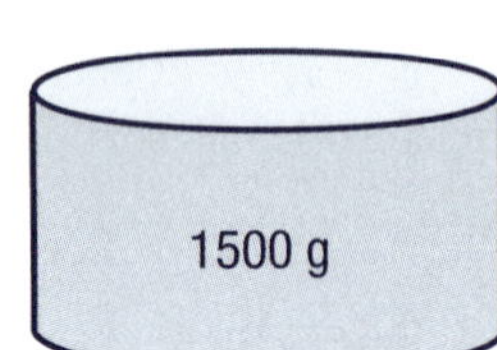

f 

Help Box

Method 2: Converting fractions to decimals

Some fractions have denominators other than 10, 100 or 1000 that are still easy to change to decimals.

For example: $\frac{2}{5} = \frac{4}{10} = 0.4$ $\qquad \frac{11}{25} = \frac{44}{100} = 0.44$

Method 3: Converting fractions to decimals

However, there are other fractions such as $\frac{5}{6}, \frac{3}{7}, \frac{2}{9}$ and so on, that cannot be converted by the two methods shown previously.

To change fractions like these to decimals we divide the numerator by the denominator.

For example: $\frac{2}{3} = 2 \div 3$

Step 1:

$$3\overline{)2} \quad \text{quotient } 0$$

Step 2:

$$3\overline{)20} \quad \text{quotient } 0.6$$

Step 3:

$$\begin{array}{r} 0.6 \\ 3\overline{)200} \\ -18\downarrow \\ \hline 20 \end{array}$$

Step 4:

$$\begin{array}{r} 0.66 \\ 3\overline{)200} \\ -18 \\ \hline 20 \\ -18 \\ \hline 20 \end{array}$$

Therefore, $\frac{2}{3}$ changed to a decimal is 0.66 (repeating).

5 Write these common fractions as decimals. (Use your calculator if needed.)

a $\frac{3}{7}$ b $\frac{4}{9}$ c $\frac{5}{6}$ d $\frac{2}{3}$ e $\frac{7}{12}$ f $\frac{5}{12}$

6 A man catches four fish that weigh: 250 g, $\frac{1}{2}$ kg, 1100 g and 1.75 kg.

a What is the total weight of the fish he caught? Write your answer as a decimal number.

b If he gave away the smallest fish, what quantity of fish would he have left? Write your answer as a decimal number.

7 A woman buys 2.3 kg of rice, $1\frac{1}{4}$ kg of fish and 200 g of salt and puts them into her empty bag. What is the total weight of the contents of her bag? Write your answer as a decimal number.

Help Box

Changing fractions and decimals to percentages

'Percentage' means 'out of 100'.

A percentage is therefore a fraction with a denominator of 100.

For example:
$\frac{34}{100} = 34\%$; $\frac{8}{10} = \frac{80}{100} = 80\%$; $\frac{1}{4} = \frac{25}{100} = 25\%$

There is a relationship between fractions, decimals and percentages.

For example:
$\frac{27}{100} = 0.27 = 27\%$; $\frac{4}{5} = 0.8 = 80\%$; $\frac{11}{20} = 0.55 = 55\%$

8 Convert these fractions and decimals to percentages.

a $\frac{3}{10}$
b 0.59
c $\frac{25}{100}$
d 0.63
e $\frac{5}{100}$
f $\frac{2}{5}$
g 1.3

9 Convert each percentage to a decimal and a fraction.

a 97%
b 39%
c 5%
d 18%
e 100%
f 25%
g 115%

10 Copy and complete the chart below to show that you understand how fractions, decimals and percentages relate to each other.

Fraction	$\frac{3}{10}$	$\frac{3}{25}$		$\frac{1}{8}$		$1\frac{1}{4}$	$2\frac{3}{5}$		3
Decimal			0.25		0.8			1.3	
Percentage									

11 A carton is packed with 750 g of flour, 1 kg of rice, 500 g of sugar and 500 g of bananas. When empty, the carton weighs 420 g. What is the total weight in kg of the packed carton? Write your answer as a decimal number and as a mixed number.

12 Write a similar problem to the example in question 11 and show how you would work out the answer by using your understanding of fractions and decimals.

Challenge

Lillian was working out the percentage of her income that she spends on food each week. She earns K250 a week.

This is what she estimated: bills = $\frac{1}{2}$, entertainment = $\frac{1}{4}$, clothes = $\frac{1}{10}$, the rest = food.

What percentage of her income does Lillian spend on food?

What amount in kina does she spend each week on each of the four items listed?

Lesson 10 Solving food problems

Recipes usually list the ingredients needed in grams, kilograms, millilitres or litres. Some recipes also include informal measurements such as 'a cupful', 'a teaspoonful' or 'a pinch of'.

1 What formal unit of measurement would you use to weigh the following?

- **a** one betel nut
- **b** a bag of sweet potatoes
- **c** a sack of coffee beans
- **d** a fish for dinner
- **e** a shipping container of coffee beans

2 Estimate then check the weight of the following.

- **a** one coconut
- **b** one pineapple
- **c** one peanut
- **d** one sweet potato
- **e** one egg
- **f** one pawpaw

3 Calculate the number of grams in the following items.

- **a** $\frac{1}{2}$ kg of flour
- **b** $1\frac{3}{4}$ kg of rice
- **c** $1\frac{1}{4}$ kg of mangoes
- **d** $2\frac{1}{8}$ kg of nuts
- **e** $2\frac{2}{5}$ kg of coffee
- **f** $\frac{2}{3}$ kg of yams

4 Find the weight of the following items in formal units of measurement:

- **a** one cup of sugar
- **b** one teaspoon of salt
- **c** one handful of peanuts
- **d** one pinch of salt
- **e** one dessertspoon of water

5 Steven started with $2\frac{2}{3}$ cups of sugar. After using sugar for cooking he was left with $1\frac{2}{5}$ cups of sugar. What fraction of the cups of sugar had Steven used?

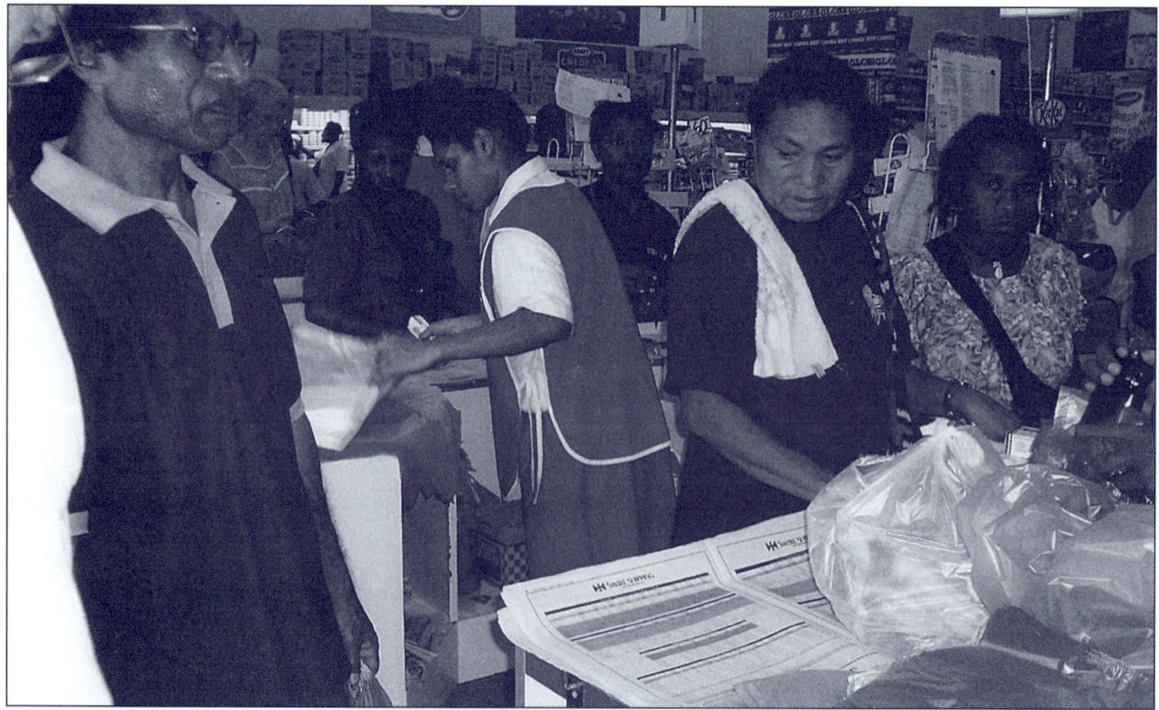

6 Annette bought $\frac{5}{8}$ kg of rice. She added it to the 1.75 kg of rice that she already had. How much rice does Annette have now?

7 Sam bought seven tins each containing 230 g of fish. How many kg of fish was in the tins altogether? Write your answer as a decimal number.

8 Write your own question about food that involves using a mixture of fractions and decimals. Work out the answer to your question.

9 The recipe said that each loaf needed $\frac{1}{5}$ cup of milk powder. How many loaves could be made from $1\frac{1}{2}$ cups of milk powder?

10 A family spends 20% of their income on vegetables, $\frac{1}{4}$ of their income on meat and $\frac{1}{5}$ of their income on fruit. What percentage of their income remains after buying these items?

11 The same biscuits were being sold in three different shops. Shop A had them on sale at 15% off the normal price of K1.70. Shop B was selling them for K1.50 a packet and Shop C usually sold them for K1.89, but had marked everything down by $\frac{1}{5}$.

a What was the sale price of the biscuits at Shop A?

b What was the sale price of the biscuits at Shop C?

c Which of the three shops has the biscuits at the lowest price?

Learning Unit 2 Healthy Living

Strand: Chance and Data

Statistics	Outcome 7.4.1	Compare sets of data
Sets	Outcome 7.4.2	Use a variety of classification methods

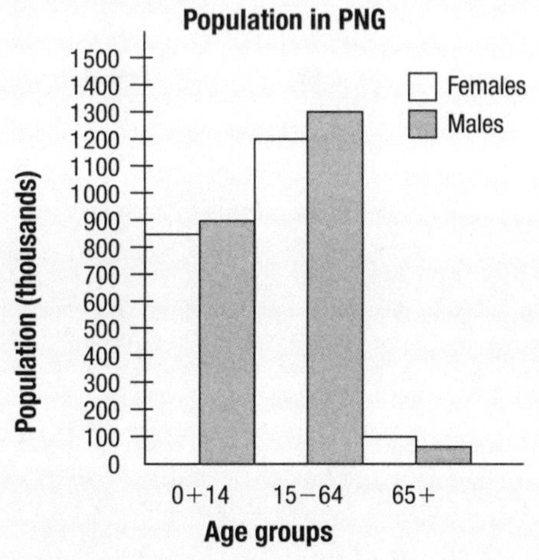

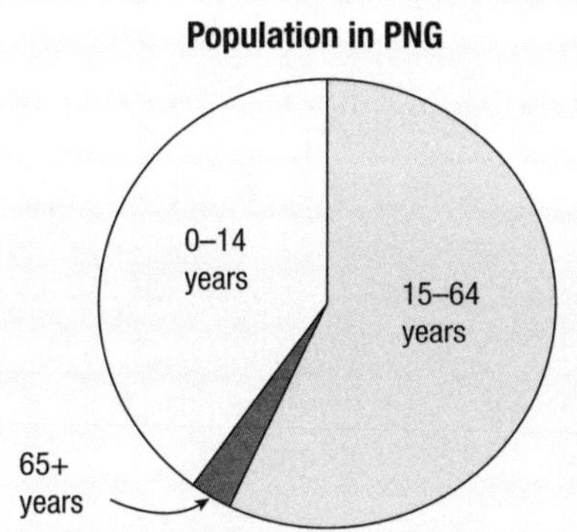

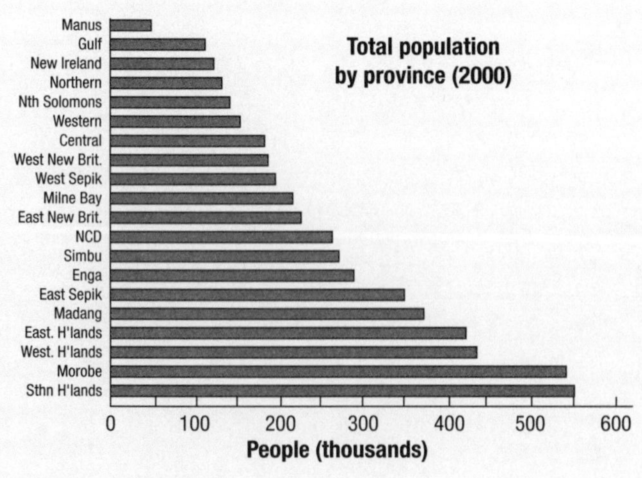

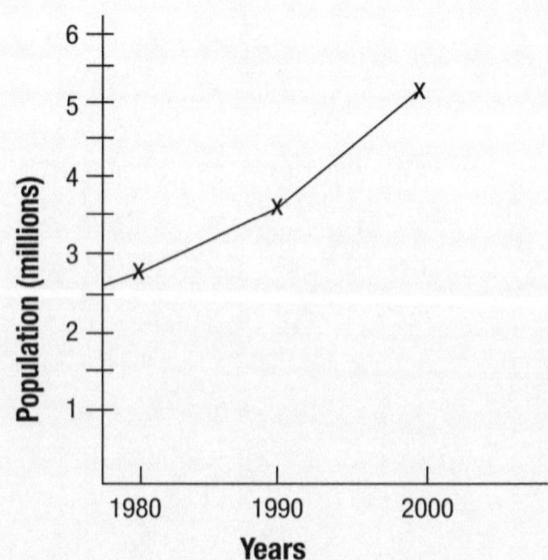

Lesson 1: Introduction	Looking at factors that affect our health
Lesson 2: What do we do to stay healthy?	Reading and drawing Venn diagrams
Lesson 3: How statistics can help	Interpreting data displayed in a range of graphs
Lesson 4: Pictograms and frequency tables	Interpreting and drawing pictograms Using a frequency table to create a bar graph
Lesson 5: Different types of bar graphs	Interpreting and drawing bar graphs and stacked bar graphs Interpreting comparative bar graphs and back-to-back bar graphs Interpreting and drawing histograms
Lesson 6: Reading line graphs and looking at scales	Interpreting line graphs Choosing suitable scales
Lesson 7: Pie or circle graphs	Interpreting and drawing pie graphs Calculating fractions to determine sectors
Lesson 8: Tree diagrams	Interpreting a range of tree diagrams Drawing tree diagrams
Lesson 9: Stem and leaf plots	Reading and creating stem and leaf plots
Lesson 10: Range, mean, median and mode	Calculating spread and measures of central tendency Relating measures of central tendency to graphs and stem and leaf plots

Lesson 1 Introduction

In this unit you will look at various statistics and graphs to learn more about the general health of Papua New Guinea's population and different ways to stay healthy.

While government health workers and local healers work to keep the community well, it is important that people also try to live a healthy lifestyle. A healthy lifestyle means getting plenty of exercise, eating a balanced diet and avoiding unnecessary risks. It is much easier to prevent ill-health than it is to cure ill-health.

These pictures show some of the different ways that people try to prevent becoming unwell.

1 Look at the pictures and describe what is happening to prevent ill health.

2 Which of these methods have you used?

3 Which of these methods do you think is the easiest for people to do to help them stay healthy?

4 What other methods can you think of for staying healthy?

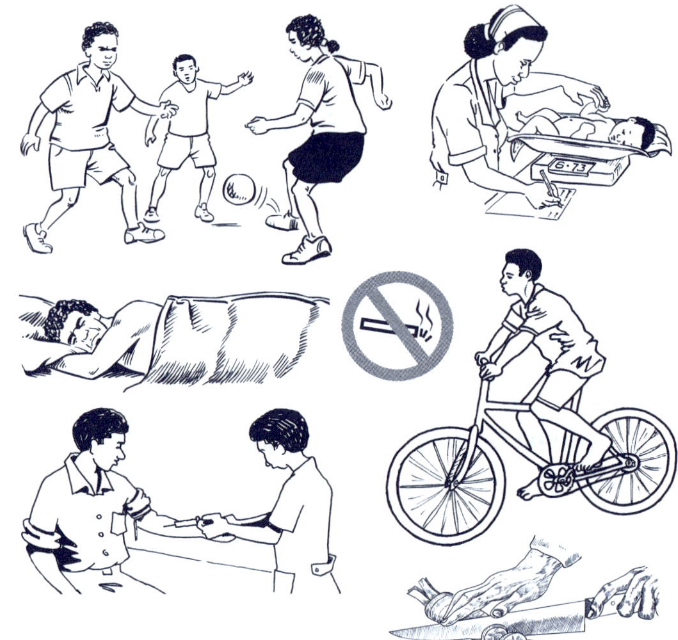

This table shows how some activities contribute to good health while others don't.

Factors affecting a populations' health	
Positive impact	**Negative impact**
exercise	smoking
healthy diet	lack of exercise
adequate rest each day	overeating
immunisation	chewing betel nuts

5 Explain why each of the items in the positive impact column is good for our health.

6 Explain why each of the items in the negative impact column is bad for our health.

7 Copy the table into your book and add a few more items to each column.

Lesson 2 What do we do to stay healthy?

This Venn diagram shows which of two sports the students in a class like to play.

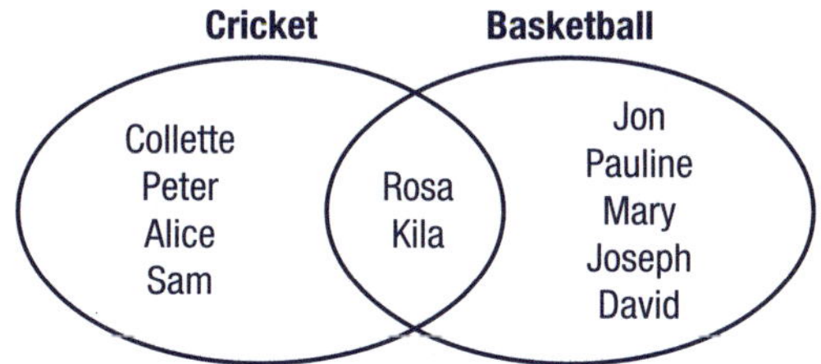

1 a How many students are in the class?

b What sport does Sam like?

c Which students like both sports?

2 Draw a Venn diagram using the following data:

- adults who swim: Lilian, Rachael, Wesley, Rosa
- adults who don't do any sport: Miro, Peter
- adults who play volleyball: Paul, Elsie, Alice, Jenny
- adults who swim and play volleyball: Mirou, Marlon

3 Copy the Venn diagram below into your book. Ask twelve students to say which of the three healthy things they do well and place the information in the correct place on the Venn diagram.

4 a Work with a friend and each list 15 things you have eaten in the past two days. Draw a Venn diagram with two intersecting circles to show the relationship between what you and your friend ate.

b What title would you give to the food in the intersecting set?

Harold drew a Venn diagram to show the set of whole numbers less than 20. He included some subsets.

Subset B is the set of even numbers less than 14 contained within A, the larger set.

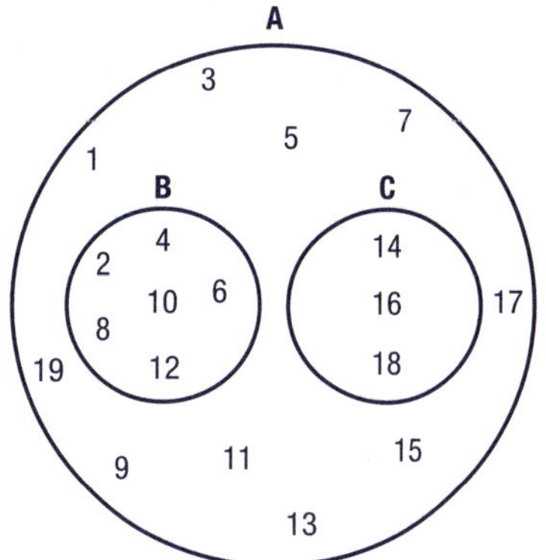

A: set of whole numbers 1 – 19
B: set of even numbers less than 14

5 a How would you describe the numbers in subset C?

b Describe the subset of numbers that are not contained in subsets B and C.

6 Draw a Venn diagram to show the set of students in your class and the subset of students who belong to a sports team.

7 Use Venn diagrams to help solve these questions:

a The nurse has to immunise 25 babies. She needs to give 10 injections against diphtheria and 20 injections against tetanus. How many babies will receive two injections?

b Twenty people belong to a sports club. Of these people, 12 like to play touch rugby and 15 like to run. Draw a Venn diagram to show how many like to do both.

c At the pool there are 8 people who belong to a soccer club and 11 people who enjoy swimming. How many people might there be in the pool? List four different answers using Venn diagrams to describe your thinking.

Lesson 3 How statistics can help

Many governments and organisations collect statistics about the population. Governments look for trends and use these to plan new facilities and services, such as schools, hospitals and roads. Companies often survey people to find out the best way to advertise and sell their products.

Data can be displayed in tables and in many different types of graphs and diagrams.

1 Name each type of graph shown below.

2 Write a few sentences about the information displayed in each graph.

3 Explain to a friend the difference between each type of graph.

a

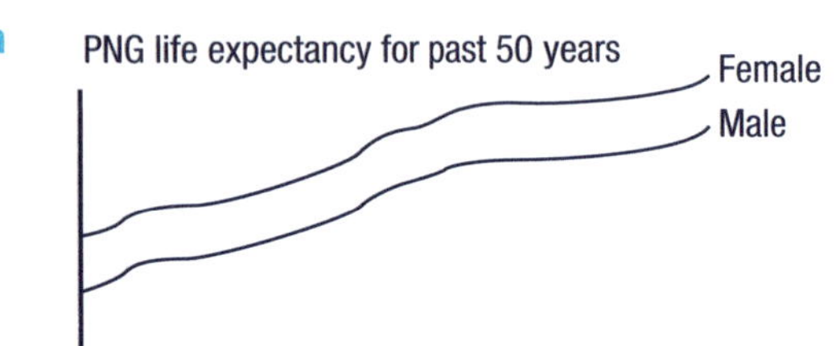

b

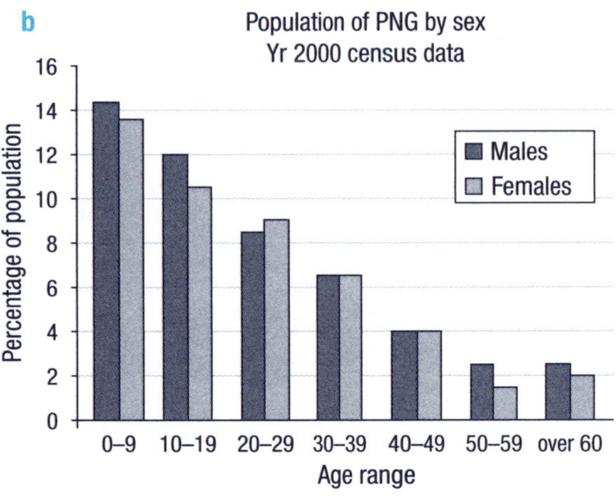

c

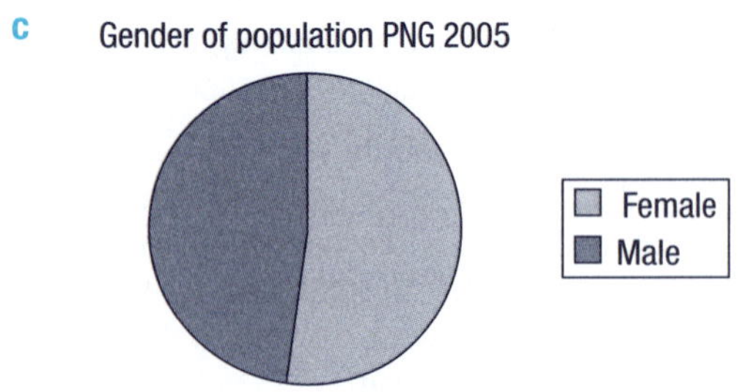

d

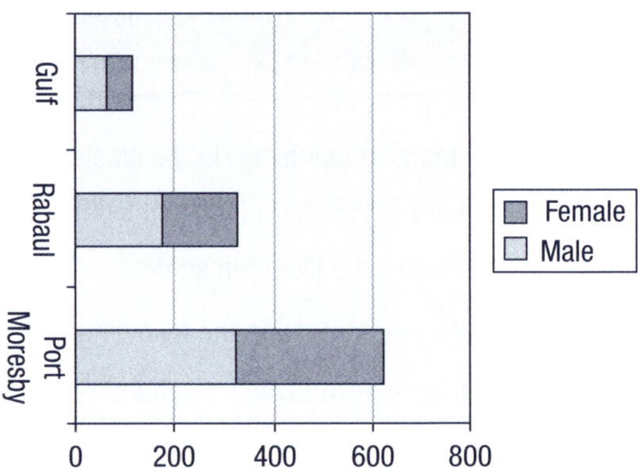

4 How might the government use the information provided in each graph?

5 Discuss how a government might improve the health of its population if it collected information about:

- a reasons people were admitted to hospital
- b types of sport played by a community
- c number of children under five years in a community
- d reasons for student absences from school

6 Survey your friends or some people in your village about a health issue and make a graph to display your data.

7 Suggest ways that this data could be used to improve the health of the group.

Lesson 4 Pictograms and frequency tables

Each symbol in a pictogram represents a certain amount. The amount is given in the legend on the graph.

1 Use the pictogram below to answer these questions:

Students	Time spent walking to school daily
Mirou	
Annette	
Keti	
Dorcas	

Legend = 10 minutes

a What is the value of one clock in this graph?

b Who spent the most time walking to school daily?

c How much longer than Annette did Dorcas spend walking to school daily?

d Which student spent 20 minutes walking to school daily?

2 a Copy the pictogram into your book and add your name and two of your friends to the graph and include the data for each person.

b Write three questions you could ask a friend about this extra data. Write the answers to your questions.

3 This pictogram shows some of the fruit students ate for lunch one week. Each symbol represents one piece of fruit.

Name	Fruit eaten at lunch for the week
Harold	
Ivy	
Alice	
Joseph	
Benti	

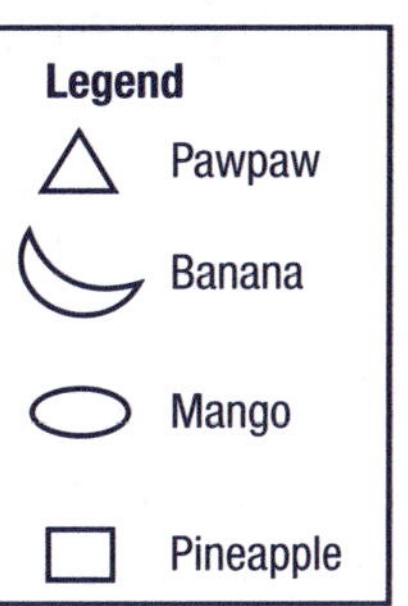

a How many different types of fruit are shown on the graph?

b Which fruit was eaten by every student listed?

c Which of the fruits listed did Harold eat the most?

d How many bananas did Alice eat?

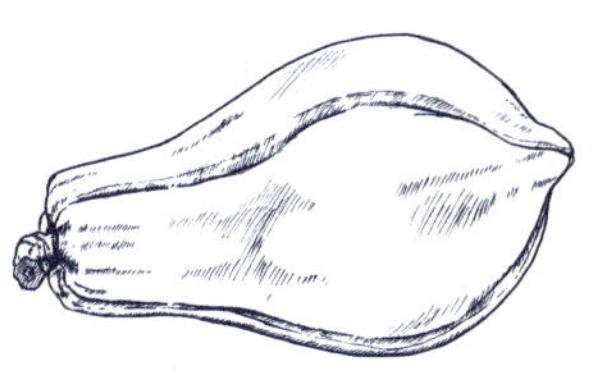

4 A health worker kept this tally of the number of immunisation injections she gave in a week.

Injection	Tally	Frequency
diphtheria	~~IIII~~ ~~IIII~~ ~~IIII~~ ~~IIII~~ ~~IIII~~ ~~IIII~~ ~~IIII~~ ~~IIII~~ III	43
measles	~~IIII~~ ~~IIII~~ II	12
polio	~~IIII~~	5
pigbel	II	2
TB	~~IIII~~ ~~IIII~~ II	12
tetanus	~~IIII~~ ~~IIII~~ ~~IIII~~ ~~IIII~~ I	21

a Use the information in the frequency table to construct a bar graph.

b What will be the title of the graph?

c What labels will be used on the x- and y-axes?

d Write three questions that could be answered by reading the graph.

5 Conduct a survey among your friends about what they do to stay healthy and create a frequency table showing the results. Graph the data in a pictogram and write a few sentences to describe what you found.

Lesson 5 Different types of bar graphs

Bar graphs use horizontal or vertical bars to display data. Another name for bar graphs that display data vertically is column graphs. The bars should be the same width.

1 The bar graph on the right shows the percentage of Papua New Guinea's population who practise different religions.

a Which religion has the most followers?

b Which religion has the least followers?

c What percentage of the population belongs to the Lutheran Church?

d What does the bar 'Other' mean?

e What percentage of the population does not belong to any religion?

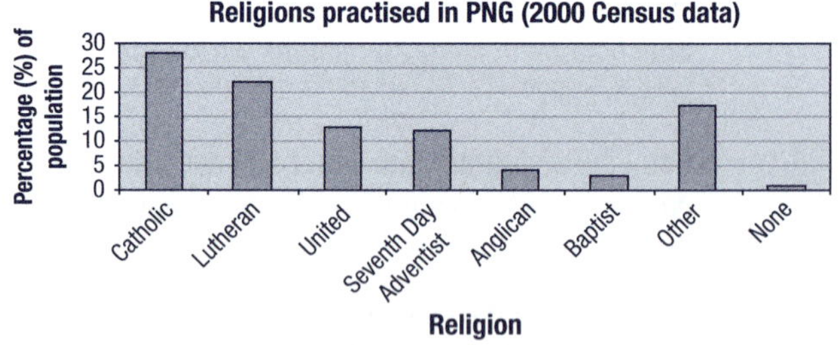

2 One hundred people were surveyed about their eating habits. List five questions they may have been asked that could have led to this bar graph being drawn.

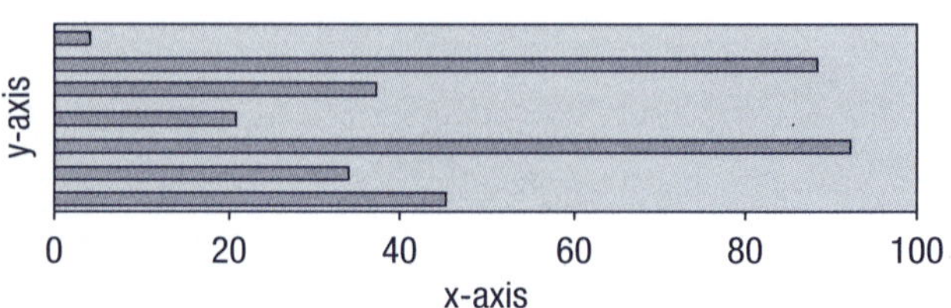

3 Based on your answer to question 2, write a title for the graph and label each axis.

4 Sometimes data is displayed in stacked bar graphs. This stacked bar graph shows both male and female information in the same bar.

- a Which age group has the largest population?
- b How many males in the 0-14 year age group?

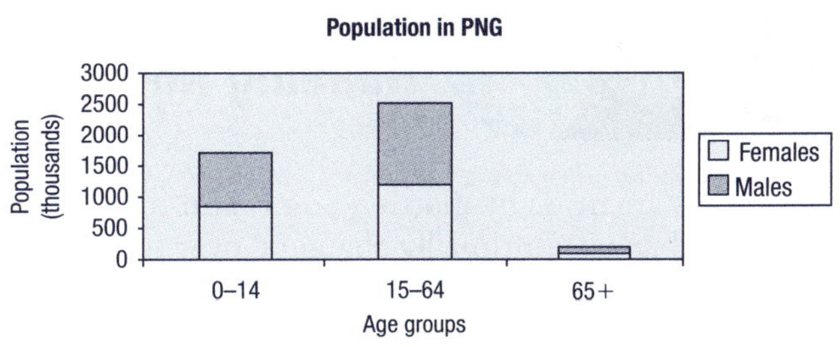

5 Collect data about the favourite sports of the males and females in your class and display your data in a stacked bar graph.

6 Bar graphs sometimes combine two or more sets of data on the one graph. This allows people reading the graph to more easily compare sets of data. The graph on the right is a comparative bar graph.

- a Write a few sentences to describe the information contained in this graph.
- b Copy the graph into your book and add your prediction for what the graph might be in 2010.
- c Explain how you decided on your predictions for 2010.

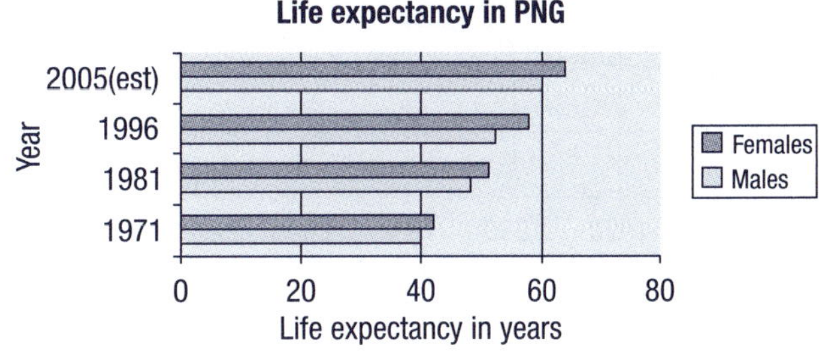

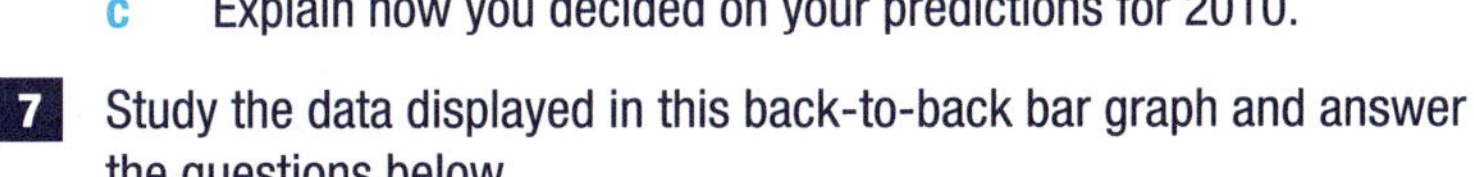

7 Study the data displayed in this back-to-back bar graph and answer the questions below.

- a The largest percentage of the population of this village is in which age range?
- b Are there more females or males in the 50+ age range for this village?
- c Describe the age spread of the population in this village according to this graph.

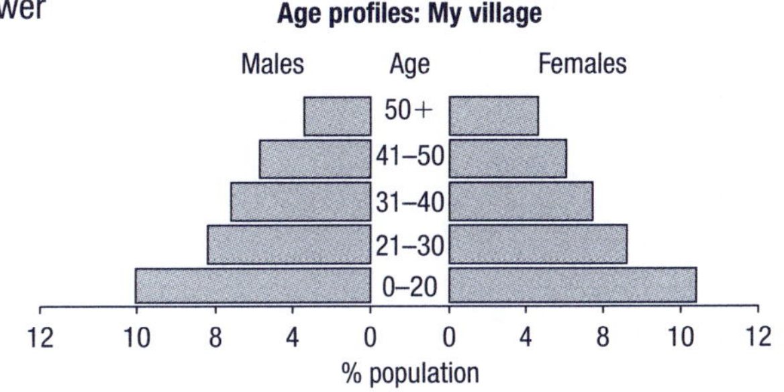

8 A histogram is another type of bar graph. The bars in a histogram touch each other and each bar relates to a category of data. The categories do not overlap at all. For example: the histogram to the right shows the number of children at a picnic within certain age categories.

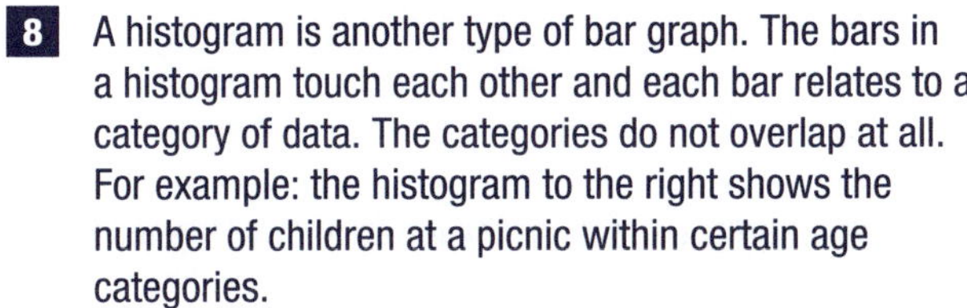

- a The majority of students at the picnic came from which age category?
- b Which age categories had more than twenty children attending the picnic?
- c How many children attended the picnic?

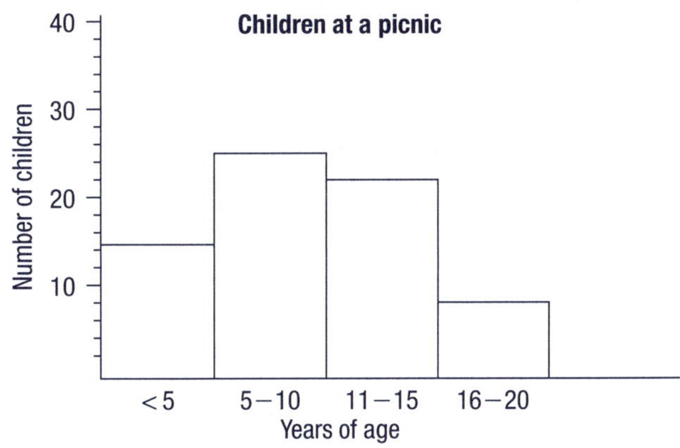

9

- a Draw a histogram to show how much time each day the students in your class spend walking to and from school. The categories could be < 5 minutes, 5–10 minutes, and so on.
- b Write three facts that can be obtained by reading your histogram.

Lesson 6 Reading line graphs and looking at scales

Line graphs are made by plotting points and then drawing line segments to join the points. Line graphs show quantities that are continually changing over time. A value can be read anywhere along the graphed line.

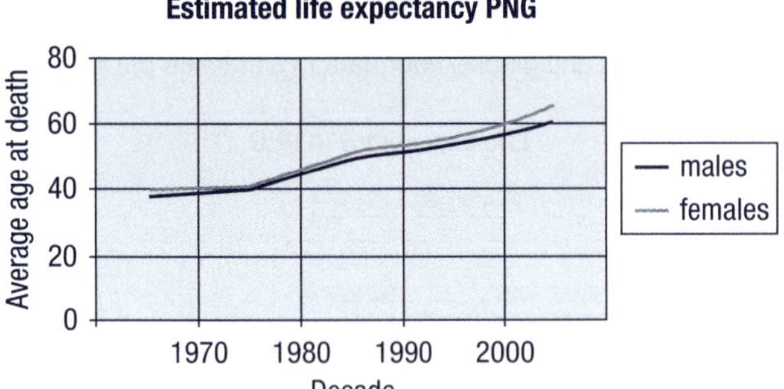

1 Look at the line graph on the left to answer the following questions.

- **a** How has the life expectancy of males changed in Papua New Guinea since 1970?
- **b** What is the life expectancy of Papua New Guinea's females in 2000?
- **c** What would you predict the life expectancy of both males and females in Papua New Guinea will be in 2010?
- **d** Why do you think the life expectancy of people in Papua New Guinea is increasing over time?

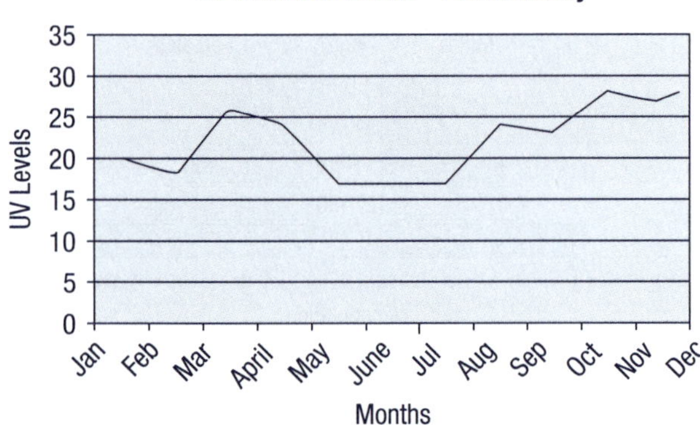

2 Look at the line graph on the left to answer the following questions.

- **a** What is the graph about?
- **b** What does the x-axis show?
- **c** In which month are the UV rays the strongest?
- **d** In which months are people safest from UV rays?
- **e** How might this data be useful?

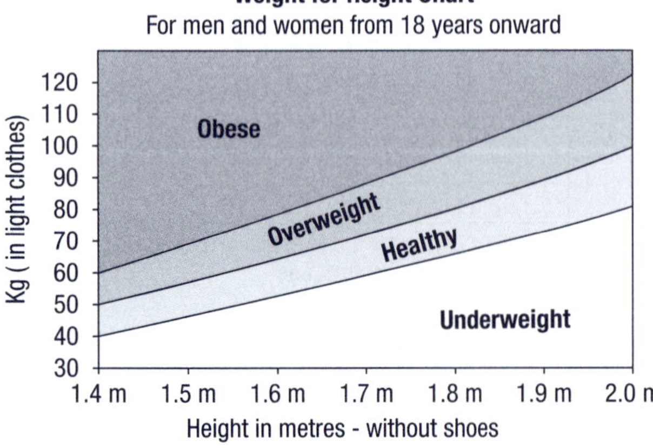

3 Look at the line graph on the left to answer the following questions.

- **a** Who might use this chart?
- **b** Into which category does a person who is 1.6 m tall and weighing 60 kg fit?
- **c** What is a healthy weight for a person who is 1.9 metres tall?
- **d** According to this chart, what weight would a 1.4 m tall person need to be to be considered underweight?

The scale used to draw a graph needs to be carefully selected to make sure that the reader is not misled by the data being displayed. Look carefully at these graphs and answer the related questions.

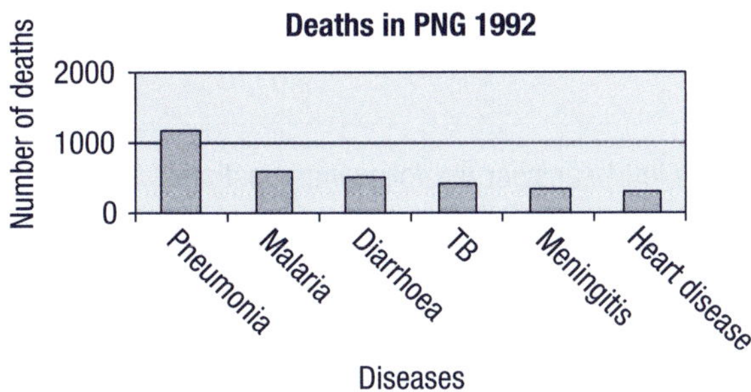

4 Look at the graph on the left to answer the following questions.

- **a** What is the graph about?
- **b** How many deaths were caused by tuberculosis (TB)?
- **c** How many more deaths were caused by malaria than by TB?
- **d** What was the main cause of death in Papua New Guinea in 1992?

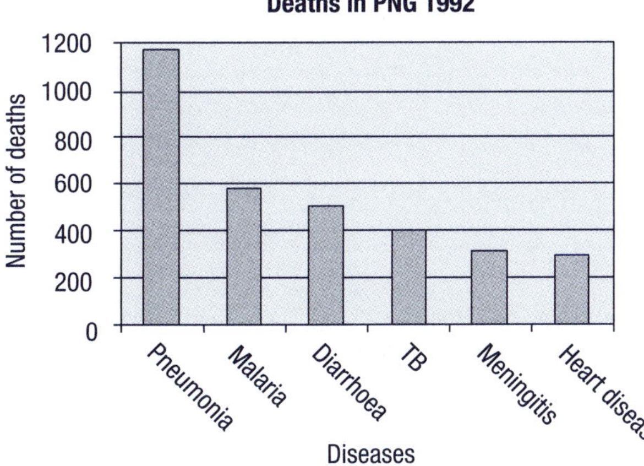

5 Look at the graph on the left and the graph used in question 4 to answer the following questions.

- **a** Does this graph show the same data as the other graph?
- **b** Which graph appears to show a greater difference between deaths from tuberculosis and deaths from malaria?
- **c** Which graph would you use if you were trying to get funding to help prevent deaths from pneumonia?
- **d** Why are the two graphs different?
- **e** Could people be misled by the first graph? Why?

6 David wanted to swim in the Pacific Games. He tried to do some swimming practice every day. He kept data about the total kilometres he swam each week for a six week period.

Week 1	Week 2	Week 3	Week 4	Week 5	Week 6
13.2 km	15.6 km	9 km	17.3 km	11.8 km	21.5 km

- **a** Select a suitable scale and draw a bar graph to represent the data in the table.
- **b** Redraw the graph so that it uses the same data but could mislead the reader.

Lesson 7 Pie or circle graphs

Pie graphs or pie charts represent data as sectors of a circle. Each sector is proportional to the overall amount of data being displayed.

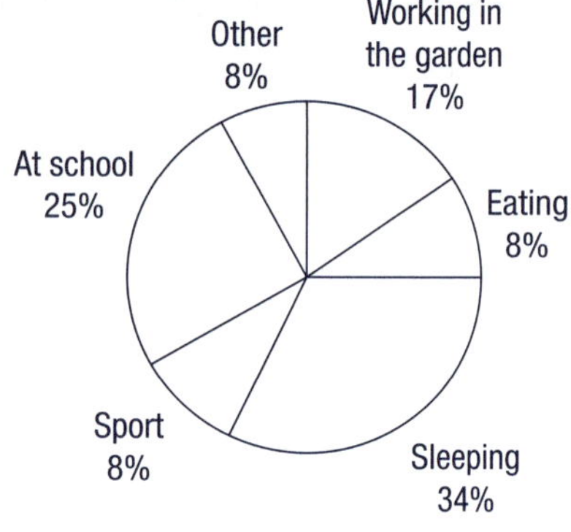

1 Look at the pie graph on the left to answer the following questions.

- **a** What does the graph describe?
- **b** What does 'Sport 8%' mean?
- **c** What fraction of the day is spent working in the garden?
- **d** How many hours are spent at school?
- **e** What angle is at the centre of the circle for the 'At school' sector?

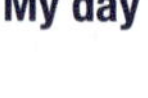

Help Box

Pie charts or graphs can be created in a number of ways. A simple method is to use your knowledge of division and fractions to decide how much of the pie will be used for each piece of data.

Example: To make a pie graph of the data shown here, follow the steps.

A = 9
B = 3
C = 4

Step 1: Total the number of units to be represented (9 + 3 + 4) = 16.

Step 2: Draw a circle and mark the centre point. Cut a piece of string the same size as the circumference of the circle.

Step 3: Mark the string into 16 equal sections. (This can be done by folding and/or measuring.)

Step 4: Mark the circle to show each of the 16 sections.

Step 5: Divide the circle into 9, 3 and 4 as shown.

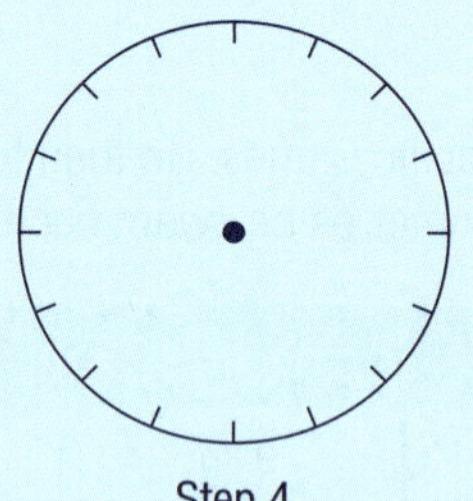

Step 4

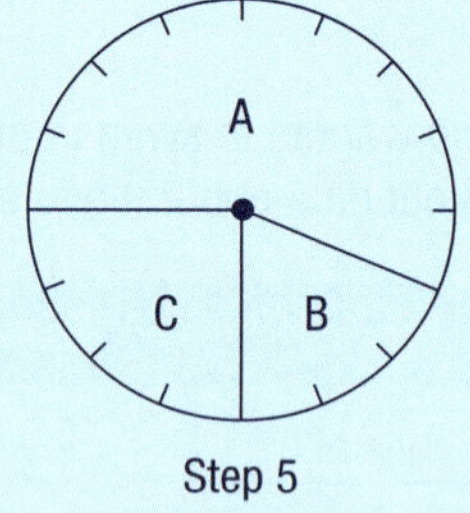

Step 5

2 Make a pie graph of the following data using the string method described above.

Fruit portions eaten by Joseph in one week				
Banana	**Pineapple**	**Pawpaw**	**Guava**	**Mango**
7	2	4	2	5

Help Box

Another method used to create a pie graph is by using degrees.

The following table shows how to calculate how much of the pie will be used for each piece of data using angles. A circle is created by turning 360° around a centre point.

Membership of sports team – Grade 7			
Sport	**Number of students**	**Fraction of total**	**Angle of sector in pie graph**
athletics	6	$\frac{6}{36}$	$\frac{6}{36}$ of 360° = 60°
soccer	8	$\frac{8}{36}$	$\frac{8}{36}$ of 360° = 80°
volleyball	4	$\frac{4}{36}$	$\frac{4}{36}$ of 360° = 40°
basketball	12	$\frac{12}{36}$	$\frac{12}{36}$ of 360° = 120°
cricket	6	$\frac{6}{36}$	$\frac{6}{36}$ of 360° = 60°
TOTAL	**36**	$\frac{36}{36}$	**360°**

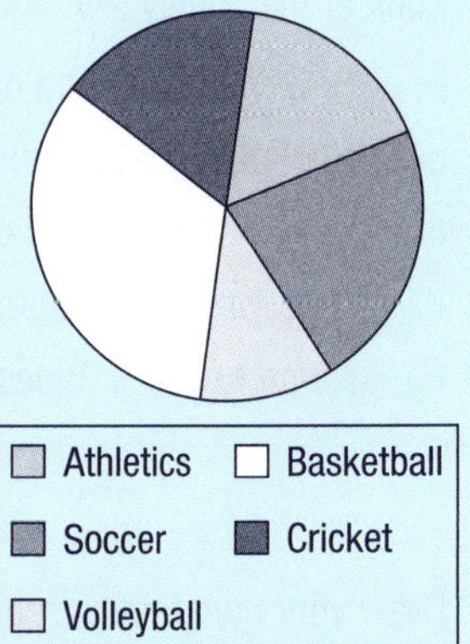

3 Copy the table below into your book and complete by working out the missing angles.

Infant immunisation program			
Vaccine	**Number of babies immunised**	**Fraction of total**	**Angle of sector in pie graph**
diphtheria	70	$\frac{70}{200}$	$\frac{70}{200}$ of 360° = °
measles	50		$\frac{}{200}$ of 360° = °
polio	30		$\frac{}{200}$ of 360° = °
TB	15		$\frac{}{200}$ of 360° = °
pigbel	35		$\frac{}{200}$ of 360° = °
TOTAL	**200**	$\frac{200}{200}$	**360°**

Challenge

Using your knowledge of angles, draw a pie graph of the data in question 3 using an estimation of the angles involved.

Write three questions that someone could answer by reading the pie graph you have drawn.

Lesson 8 Tree diagrams

These diagrams are an easy way to show the different ways that information can be linked. A 'family tree' is an example of a tree diagram.

1 Look at this family tree and answer the questions below.

- a How many children do Grandma Ivy and Grandpa Jon Lokei have?
- b How many of their children are married?
- c How many grandchildren do they have?
- d How many great grandchildren do they have?
- e Name two of K Rapese's cousins.

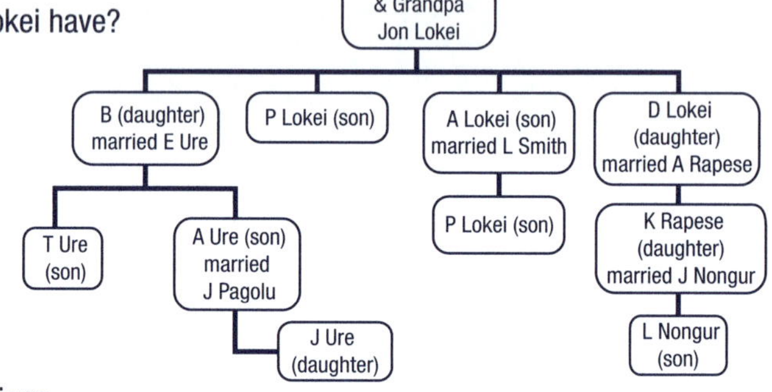

2 Draw your own family tree showing between 3 and 5 generations.

3 The nutrition information on a packet of biscuits is set out like this:

- a What are the three main nutritional components of these biscuits?
- b List the different types of fat in these biscuits.
- c According to the diagram what are the minor items that make up the 19.7 g of carbohydrates?

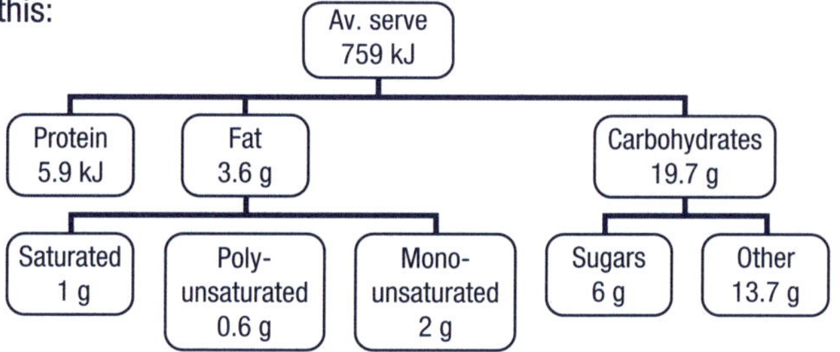

4 Naomi has a few different choices for her lunch. She can have a sandwich (S) or a roll (R) with butter (B) or margarine (M), and filled with tomato (T), tinned meat (TM) or fish (F).

- a How many different kinds of sandwiches and rolls can Naomi make? Copy the tree diagram on the right into your book and finish it to find out.
- b Draw another tree diagram to work out how many choices Naomi would have if she had a choice of two types of bread (white or wholemeal), two spreads (butter or margarine) and two fillings (banana or tinned meat).

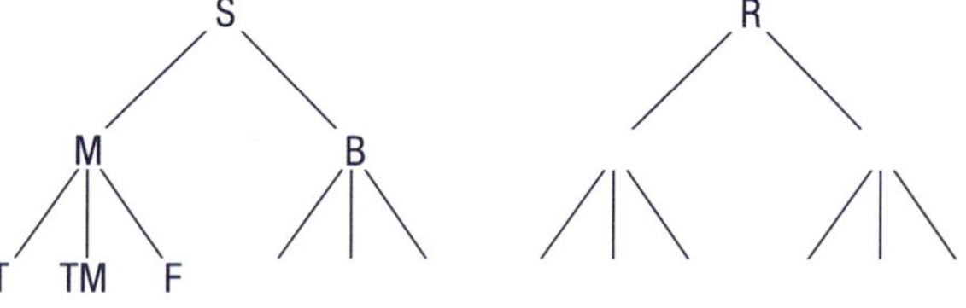

- c Draw a tree diagram to show her choices if she had to choose between three cordials (red, yellow or green), two fruits (banana or mango) and four cakes (chocolate, vanilla, lemon or strawberry).
- d What pattern can you see that might help you solve this type of problem without having to draw a tree diagram?

5 David has three different coloured shirts and three different coloured shorts.

- a How many different combinations can he make? Draw a tree diagram to prove your answer.
- b What pattern can you see that might help you solve this type of problem without having to draw a tree diagram?

6 Nat has time to be in two different sports teams on the weekend. He can choose between cricket, soccer, basketball, tennis or rugby. List all the combinations that Nat can select from.

7 a List three drinks, two kinds of meat and three vegetables that you like.

b Describe how you might work out how many different combinations of one meat, one vegetable and one drink you could have.

Lesson 9 Stem and leaf plots

Stem and leaf plots are used to display data in a way that separates the digits into two parts. For example; when comparing digits that have up to two numbers the tens column becomes the **stem** and the units column becomes the **leaves**. The digits 9, 13, 15, 13, 19 put into a stem and leaf plot would look like this:

stem	leaf	
0	9	
1	3, 3, 5, 9	**Key:** 1|3 stands for 13

Stem and leaf plots always have a key. The key to this stem and leaf plot is shown on the right.

The leaves are usually organised from lowest to highest. Only one digit from each number can be written in the leaves column, but the stem can have any number of digits.

1 Write stem and leaf plots for the following sets of numbers. Remember to include a key.

a 6, 5, 12, 19, 16, 10, 15.

b 24, 13, 6, 29, 31, 11, 16, 53, 27, 9.

2 Organise these sets of data into ordered stem and leaf plots:

a ages of people in a house
47, 41, 32, 3, 62, 7, 9, 64

b time in minutes taken by grade 7 to run once around the playground
13, 22, 15, 14, 14, 32, 10

c runs scored by a team in cricket
6, 15, 0, 11, 3, 11, 8, 10, 2, 15, 9

d price of food items in toea
39, 45, 23, 25, 15, 12, 18, 29

3 A teacher used this stem and leaf plot to record how many days illness each of her students had during the year.

stem	leaf	
0	0, 2, 3, 4, 7, 9	
1	0, 0, 1, 1, 3, 4, 5, 7	
2	0, 1, 1, 2, 2, 2, 2, 3, 5	
3	2, 3	**Key:** 3|7 stands for 37

a How many students in the class?

b What was the least number of days any student was absent due to illness?

c What was the most number of days any student was absent due to illness?

4 This stem and leaf plot shows the ages of people in a village who were seen by the health worker in one week.

stem	leaf	
0	1, 1, 1, 2, 3	
1	1, 1, 9	
2	4, 8	
3	2	
5	2, 3, 3, 6, 6, 8, 9	
6	2	**Key:** 6|2 stands for 62

a Write a list of the ages of the people who were seen by the health worker.

b How many people were seen by the health worker in the week?

c What were the ages of the youngest and eldest patients?

d What does this stem and leaf graph tell about who are the biggest users of health care?

5 Create a stem and leaf plot to show the ages of the people in your family.

Lesson 10 Range, mean, median and mode

Once data has been collected and displayed in tables or graphs it can be analysed in different ways. Three ways to measure the centre or middle of the data are to calculate the mean, the median and the mode. These are all known as 'measures of average'. Another way of measuring data is to look at the range of the data.

Help Box

The **mean** of a set of data is often called the **average**. To find the mean, total the scores in the data set and then divide by the number of scores. Mean $= \frac{\text{sum of scores}}{\text{number of scores.}}$

Example: In the data set (2,3,3,5,7) the mean is $(\frac{20}{5}) = 4$.

The **median** of a set of data is the middle score. To find the median, put the data in numerical order and find the middle one.

Example: In the data set (2,3,3,5,7) the median is 3.

If there is an even number of data the median is halfway between the two middle scores.

Example: In the data set (2,3,5,7) the median is halfway between 3 and 5. The median is 4.

The **mode** of a set of data is the most frequently appearing score.

Example: In the data set (2,3,3,5,7) the mode is 3.

The **range** is the spread between the highest and lowest scores.
Range = highest score − lowest score.

Example: In the data set (2,3,3,5,7) the range is (7–2) = 5.

The following table has some data about the height of the village basketball team.

Name	Rodney	Pauline	Raya	Jara	Leti	Sam	Dorcas	Maria
Height (cm)	162	160	156	160	159	156	163	156

1 What is the team's:

- a mean height?
- b median height?
- c mode height?
- d height range?

2 The team scored 14, 38, 26 and 18 in its last four games. What is its mean score?

3 The village team has to play against a team that has a mean score of 25 for its last four games. What might its four game scores have been?

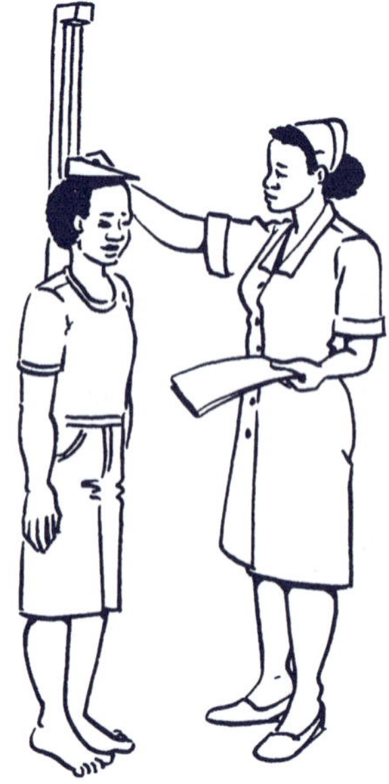

4 The stem and leaf plot below shows how many minutes it has taken Lena to swim one kilometre during training sessions over the past month.

stem	leaf
1	8, 8, 9, 9, 9
2	0, 0, 0, 1, 1, 1, 1, 3

Key: 1|8 stands for 18

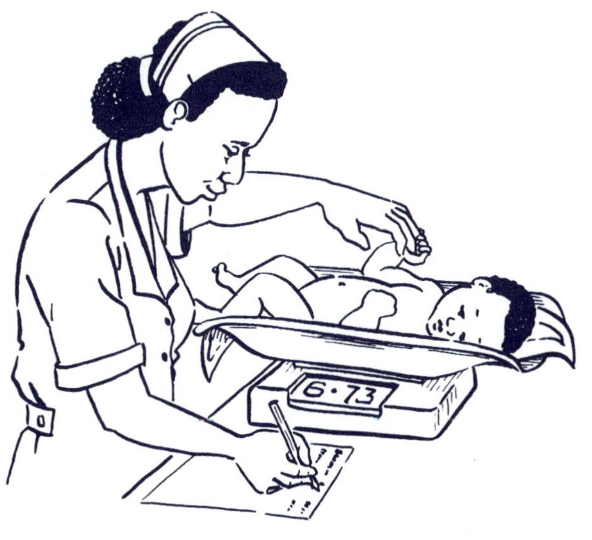

a What was Lena's mean time?

b What was her median time?

c What was her mode time?

d What was the range of her times?

e How does using a stem and leaf plot make it easier to work out the mode?

5 Five newborn babies in Port Moresby hospital have a mean weight of 3 kg. What might be the individual weight of each baby?

6 Using the weights that you listed in question 5, calculate the median and mode weights of the five babies that you described in your data set.

7 Copy this graph into your book.

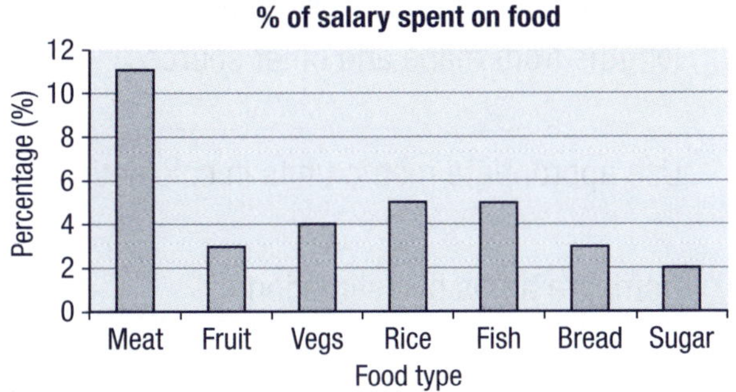

MARGARINE

a What is the graph about?

b What does the y-axis show?

c Calculate and rule lines across to show the mean, median and mode of the data displayed in the graph.

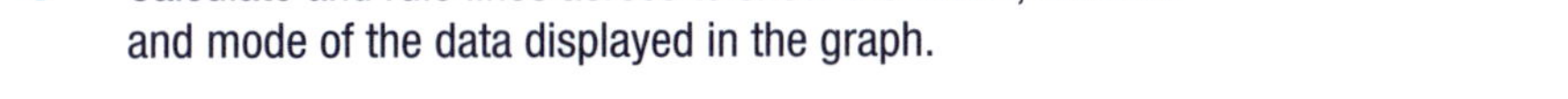

8 a Measure the height of six students in your class. Make a table and a bar graph to display your data.

b Calculate the mean, median, mode and range of the data set.

c Rule lines on your graph to show the mean, median and mode.

9 Describe a situation when:

a the mean would be the best measure of average to use

b the median would be the best measure of average to use

c the mode would be the best measure of average to use

Learning Unit 3 How Do I Compare?

Strand: Number and Application

Ratios and Rates	Outcome 7.1.6	Recognise and relate rates to graphs

Strand: Measurement

Weight	Outcome 7.3.2	Recognise weight as a force

Strand: Space and Shape

Length	Outcome 7.2.1	Estimate and measure using metric units and lengths from maps and other sources
Length	Outcome 7.2.2	Use appropriate metric units in calculations
Area	Outcome 7.2.4	Compare areas by estimation
Capacity	Outcome 7.2.7	Investigate the relationship between capacity and volume

Strand: Chance and Data

Error and Accuracy	Outcome 7.4.4	Apply strategies to reduce error
Estimation	Outcome 7.4.6	Use a variety of estimation strategies

Lesson 1: Introduction	Looking at the traditional units of measurement that used body parts Working with the metric system
Lesson 2: Reading measurement scales	Understanding and reading the graduations on different scales Choosing appropriate measurement units
Lesson 3: Weight	Estimating and converting weight units Choosing suitable weight units
Lesson 4: Length, height and distance	Estimating and converting length units Choosing suitable length units
Lesson 5: Rate	Comparing time, distance and speed Making appropriate estimates
Lesson 6: Area	Estimating, calculating and comparing areas Finding areas of compound shapes
Lesson 7: Capacity	Converting between mL and L Estimating and calculating capacity
Lesson 8: Relating capacity and volume	Converting between capacity and volume Estimating capacity and volume Using displacement to find the volume of irregular solids

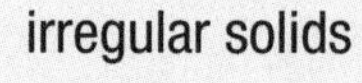

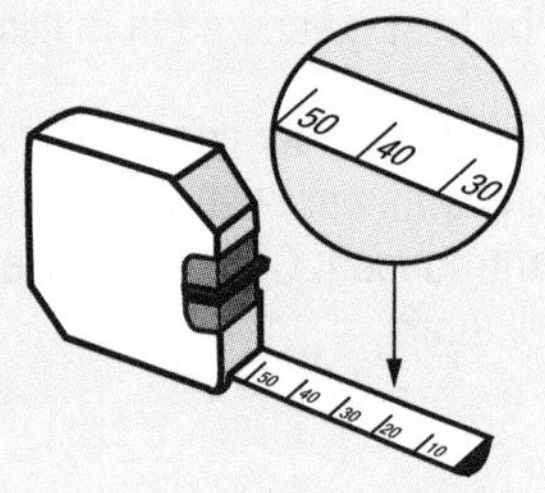

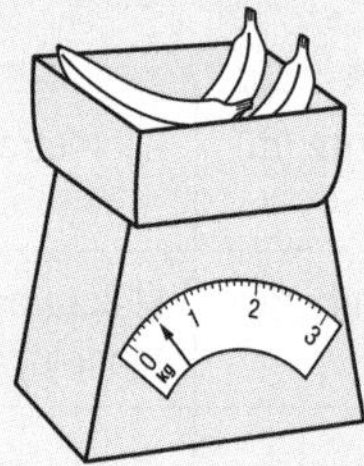

Lesson 1 Introduction

During this unit you will revise and extend some of the work you have already done with measurement including length, area, weight and time. You will also learn about your own abilities and physical attributes and how they compare to other people.

In the past, people based their measurements on parts of the body.

A foot was the measurement from the heel to the toe.

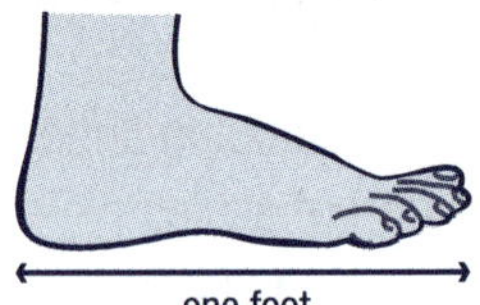

A cubit was the distance from the elbow to the middle fingertip.

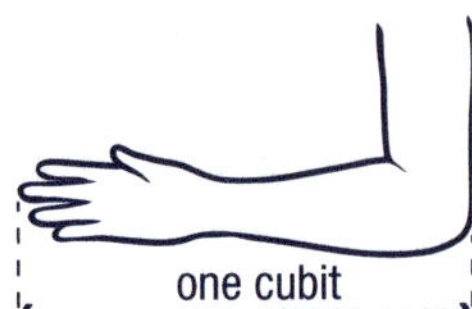

A span was the measurement from the tip of the little finger to the tip of the thumb when the hand was stretched out.

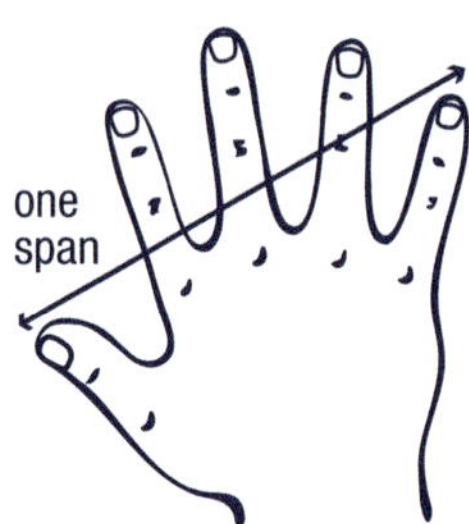

A fathom was the distance between your fingertips when your arms were stretched out sideways.

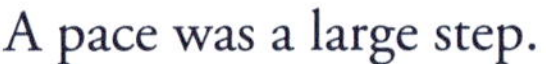
A pace was a large step.

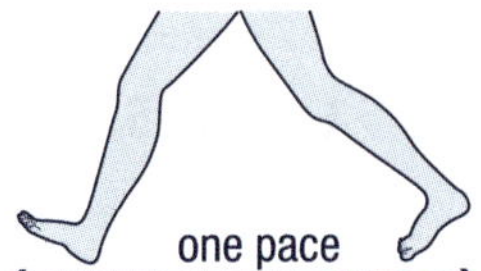

This measurement system was confusing because these lengths differ between people.

1 Use a metric ruler to measure the actual length in centimetres of:

a your foot
b your span
c your cubit
d your fathom
e your pace

Papua New Guinea now uses the **metric** measurement system. The metric system is based on tens, hundreds and thousands. Using standard units of measurements makes it easier to share information accurately because everyone interprets the measurements in the same way.

2 Copy the following table into your book.

The metric system			
	Small units	**Unit**	**Large units**
Length	millimetre (mm) centimetre (cm)	metre (m)	kilometre (km)
Weight	milligram (mg)	gram (g)	kilogram (kg) tonne (t)
Temperature	Celsius (°C)		
Capacity	millilitre (mL)	litre (L)	kilolitre (kL)

3 Describe something that would be most appropriately measured using each of the metric units listed in the table. (For example: a glass of kulau would be best measured in millilitres.)

4 Speak to the elders in your community about the traditional units that were used before the metric system was introduced. Compare these to the modern metric units.

Lesson 2 Reading measurement scales

Most families use a range of different measuring devices every day. These include rulers, measuring tapes, weighing scales, thermometers, odometers in cars, dials on the oven, clocks, measuring jugs and trundle wheels. Different measuring devices use different scales. To measure accurately, we need to understand the way each scale has been divided into major and minor units. This is called the **graduation** on the scale.

Some measuring devices do not measure right from the end, such as this ruler.

1 Read the correct measurement on each of the scales shown below.

a

b

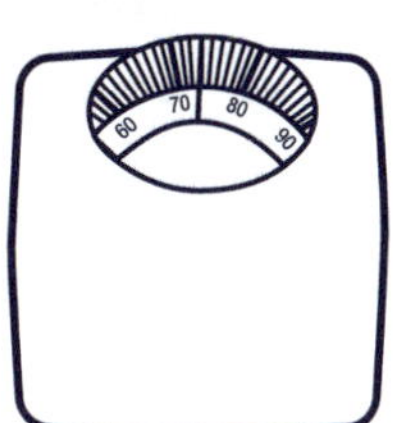

c

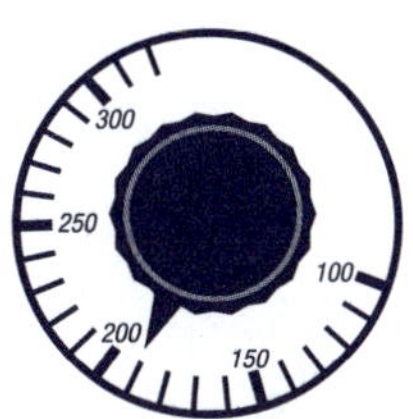

d

e

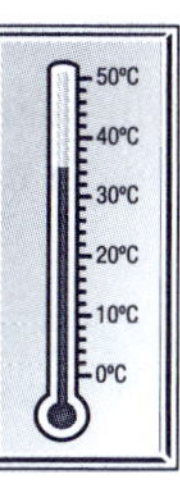

f

g

h

2 What measurement is shown on each of the following diagrams? (Include the appropriate unit of measurement in your answer.)

a

b

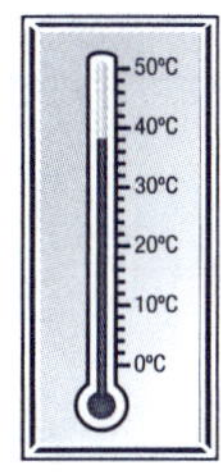

c

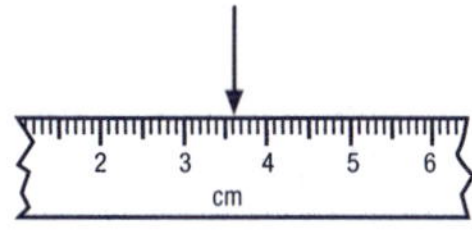

d

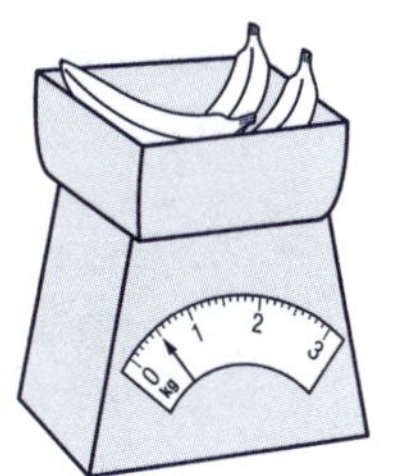

e

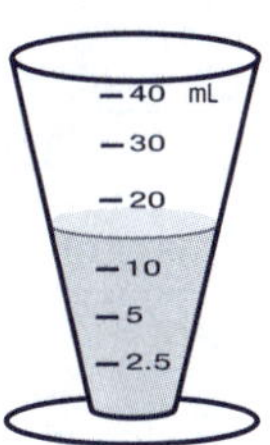

f

3 For each of the following, state what is being measured (weight, time, length …) and which unit of measurement would be the best to use (kilograms, hours, millimetres …):

- a how heavy a baby is
- b how tall a person is
- c how long it takes to run 100 metres
- d how far it is from one end of a pool to the other
- e the amount of water in a drink bottle
- f the distance around a running track
- g the angle a diver enters the pool at
- h the thickness of a fingernail
- i the heat of a drum oven

4 Draw a line 20 cm long and mark off the following lengths.

- a 5 cm
- b 8 cm
- c 19.5 cm
- d 70 mm
- e 153 mm
- f 5.75 cm
- g $6\frac{1}{10}$ cm
- h 0.12 m

5 Estimate then measure two items in the classroom that:

- a are each 15 cm long
- b have a combined weight of 1 kg
- c when laid end to end are less than 30 cm long together
- d have a combined weight of less than 300 g
- e are equal in width
- f are equal in weight

Lesson 3 Weight

The weight of an object is the measure of its downward pull. The metric system measures weight in milligrams (mg), grams (g), kilograms (kg) and tonnes (t). To estimate weight accurately it is good to have a mental picture of something that is roughly equal to each of the four units of weight.

Unit	Symbol	Example
milligram	mg	a grain of sand
gram	g	a grain of rice
kilogram	kg	one litre of water
tonne	t	14 bags of copra

1 Make a table similar to the one above and include some local items that would be roughly equal to each of the four units of weight.

2 Choose suitable units and estimate the weight of:

- a a pencil
- b your desk
- c a car tyre
- d a glass of kulau
- e your teacher
- f an eyelash
- g an aeroplane
- h a banana

3 Compare your estimates with a partner and discuss whether differences are small enough to be acceptable within the level of accuracy required for this estimation task. If you think any differences are too large, try to determine the actual weight of the object.

This spring balance works by measuring the downward pull on the spring. All weighing devices use springs, rubber bands or force meters to measure weight as a downward pull.

4 List or draw at least three other devices used for weighing objects.

5 Which weighing machine would be most appropriate for weighing:

a a fish? b a baby? c some gold dust?

d fruit at the market? e flour for a cake?

Help Box

Converting different weights to the same unit of measurement makes calculations easier.

Converting units of weight

This diagram shows that to convert large units to smaller units you **multiply** and to convert smaller units to larger ones you **divide**.

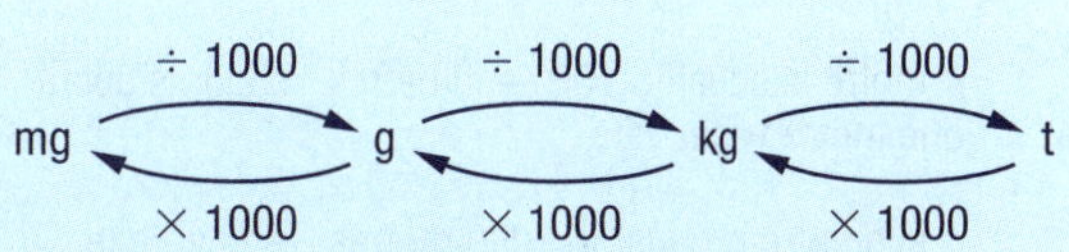

So 5000 mg is (5000 mg ÷ 1000) = 5 g

So 8 t is (8 t ×1000) = 8000 kg.

6 Copy the following statements into your book and make the conversions to fill in the spaces.

a 3000 g = ____ kg b 8 t = ____ kg c 1250 mg = ____ g

d 4750 kg = ____ t e 57 kg = ____ g f 2.5 kg = ____ g

g $4\frac{3}{4}$ t = ____ kg h $2\frac{1}{2}$ kg = ____ g i 2.25 t = ____ g

Many athletes lift weight as part of their fitness routine. These weights can be small such as 250 g carried in each hand while running or walking. At the 2006 Commonwealth Games, the weight lifter D. L. Toua successfully lifted 181 kg above his head to win a silver medal.

7 List three objects (or collections of objects) that you think might be about equal to 181 kg.

8 Fourteen bags of dry copra weigh about 1 tonne. About how many bags of copra are equal in weight to D. L. Toua's Olympic heft?

9 A small lift in a hotel can hold 800 kg.

a Work in a group to estimate the average weight of five students and their schoolbags.

b How many students with their schoolbags should be allowed in this lift at any one time? Explain your answer.

Lesson 4 Length, height and distance

Sometimes it is necessary to estimate different lengths, heights or distances. To make an accurate estimation it is a good idea to have a mental picture of something that is about the same length as the most common units of length.

A pin is about one millimetre thick.

A centimetre is equal to 10 millimetres. A fingernail is about one centimetre wide.

A metre is equal to 100 centimetres. A door is about one metre wide.

A kilometre is equal to 1000 metres. The average person can walk about one kilometre in 15 minutes.

10 mm = 1 cm
100 cm = 1 m
1000 m = 1 km

1 Choose suitable units and estimate:

- a the height of a flagpole
- b the length of your finger
- c the distance between you and the nearest haus sik
- d the distance between the school and your home
- e the thickness of a piece of cotton fabric

2 Estimate and then measure each of these lengths in millimetres and centimetres.

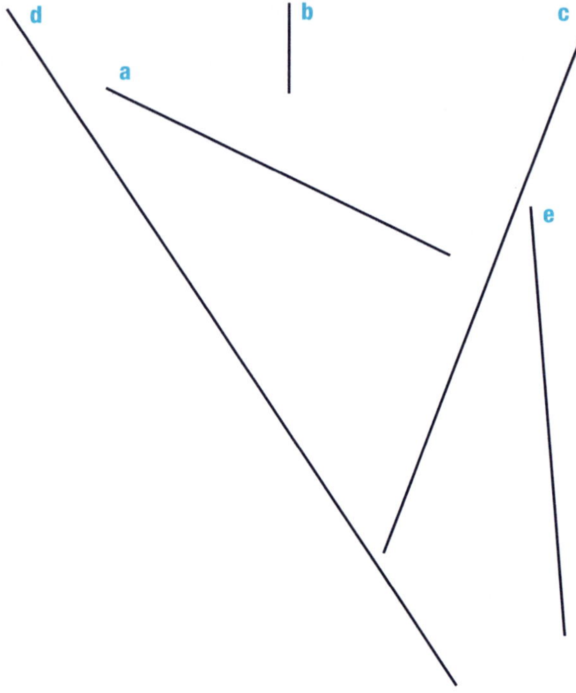

3 Estimate, then using the instructions in the Help Box below measure:

- a your height
- b the length of your normal walking stride
- c the length of your normal running stride
- d the length you can jump with your feet together from a standing position

To measure your normal walking and running strides work with a partner and follow the instructions below.

- Mark out a flat 10-metre section in the schoolyard.
- Walk the 10-metre length with a normal walking stride.
- Have your partner count how many strides you take to cover the 10-metre section.
- Repeat a few times and work out the average number of normal strides over 10 metres.
- Divide 10 metres by the number of strides to find the length of your average walking stride.

Follow the same steps to work out the length of your average running stride.

4 Collect data from four other students about their height, walking stride and running stride.

a Copy the table on the right into your book and complete it using your data and the data you collected from your friends.

b Calculate and include the mean average of each of the three measurements in the table.

c Write the students names and measurements in order from most to least for each of the four activities.

d Does there seem to be a relationship between height and walking stride?

e Does there seem to be a relationship between walking stride and running stride?

f Write a few sentences to describe some of the differences between the students' data.

g Calculate the difference between your personal data and the three averages.

Name	Height	Walking stride	Running stride
Average			

Help Box

Converting different lengths to the same unit of measurement makes calculations easier.

Converting units of length

This diagram shows that to convert large units to smaller units you **multiply** and to convert smaller units to large units you **divide**.

mm → (÷ 10) → cm → (÷ 100) → m → (÷ 1000) → km

km → (× 1000) → m → (× 100) → cm → (× 10) → mm

So 5 km = (5 km × 1000) = 5000 m.

So 6000 mm = (6000 mm ÷ 10) = 600 cm, and $(\frac{600 \text{ cm}}{100})$ = 6 m.

5 Tuan wanted to work out who had the longest running stride in his group: Tuan 940 mm, Pete 85 cm, Sera 0.82 m, Lily 87 cm and 7 mm, Luana 91.3 cm

a Convert all the measurements to centimetres.

b List group members in order from longest running stride to shortest running stride.

6 Convert the following sets of measurements to the same unit and then write each set in order from largest to smallest.

a 3.7 cm, 390 mm, 0.03 m

b 82 mm, 0.92 cm, 0.0085 m

c 0.3 km, 480 m, 4000 cm

d 0.8 km, 9.5 m, 30 000 mm

In 1988, G. Chistyakova from Russia set a world record for women's long jump by jumping 7.52 metres. Her record remains unbroken to date.

7 Measure out a distance of 7.52 metres in the school ground.

8 Estimate then measure how far you can jump doing a long jump.

9 Copy and complete the table to compare how far you jumped and how far Chistyakova jumped in each of the different units of measurement.

Name	Distance jumped in mm	Distance jumped in cm	Distance jumped in m
Chistyakova			7.52 m
Difference	mm	cm	m

10 Convert the different lengths to the same unit of measurement and calculate.

a 300 cm + 8.7 m + 0.3 km

b 24 m + 314 cm − 120 mm

c 0.2 km − 85 m + 340 cm

11 Elessala swam 50 lengths of a 25 m pool. How many km did he swim?

12 Kekeni had run 700 m of a 3 km race. How far did she still have to go?

13 The running track is 0.5 km around. How many laps will Bernadette run in a 1500 m race?

14 The school long jump record was 2.72 m. Daniel jumped 4.8 cm less than the record and Bill jumped 0.23 m more than the record. How far did each boy jump?

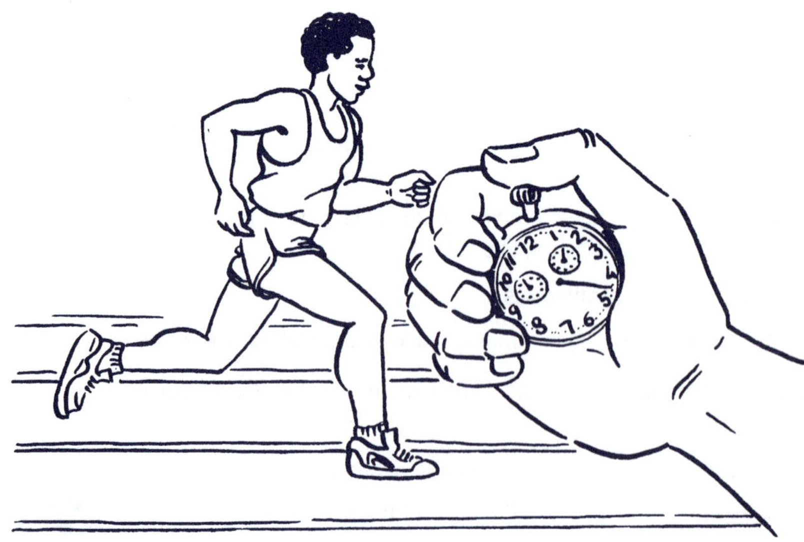

Lesson 5 Rate

In many situations people estimate how long it might take them to complete lengthy tasks based on how long it takes them to complete similar, shorter tasks. For example: if you know how long it takes you to walk one kilometre, you can estimate how long it might take you to walk five kilometres. Of course this estimation relies on you walking at the same pace for one kilometre as you would for five kilometres. This may not always happen.

Help Box

Think about how you might decide the answer to the question using the information given.

The boy cycles 6 km in 30 minutes.

How far might he cycle in 3 hours?

One way of solving this problem is to think about it as a number sentence with some parts missing.

For example: If 30 minutes = 6 km, 180 minutes is six times longer than 30 minutes, so the boy could cycle six times as far.

Then 180 minutes = ? km

Another way of solving the problem is to draw a diagram, such as the one below.

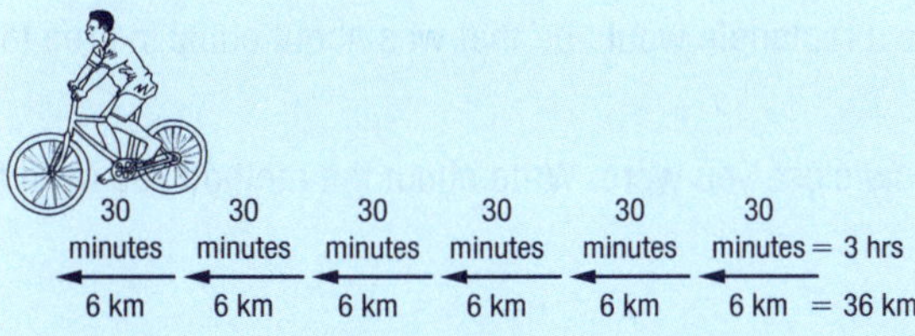

1 Measure out a 10 metre length and work with a group of students to time how long it takes each of you to walk and run that distance. Write your answers into a table.

2 Calculate how long it might take each of you to walk one kilometre, if you walked at the same speed as you did for the 10 metres. Show how you worked out your answers.

3 Calculate how long it would take each of you to run one kilometre if you ran at the same speed as you ran the 10 metres. Show how you worked out your answers.

4 Give reasons why using the 10 metre sample may not provide a very good estimate of how long it would take someone to run 20 kilometres around their own neighbourhood.

5 Calculate the following by using the information provided.

- **a** A heart beats 12 times in 6 seconds. How many beats in one hour?
- **b** David can skip rope 87 times in one minute. How many skips in three minutes?
- **c** A boy swam 250 metres in 12 minutes. How long to swim 5 kilometres?
- **d** Tom can clap 5 times in two seconds. How many claps in one minute?

6 Look at the examples given in the previous question. Which examples are likely to give a reasonable estimate for a longer time than was described? Explain why this might be.

7 Calculate the following based on the information provided.

- **a** I lapped the pool five times in 15 minutes. How long might I take to swim one lap?
- **b** I walked 10 kilometres in two hours. How long might it take me to walk one kilometre?
- **c** I jogged 18 kilometres in 1.5 hours. How far might I jog in 10 minutes?

Lesson 6 Area

Area is the measure of the amount or size of a surface. It is measured in square units such as square centimetres (cm^2) or square metres (m^2). The area is the number of square units needed to cover the surface.

1 a Find the area of each of these shapes by counting the shaded squares.

Scale 1 unit = 1 cm^2

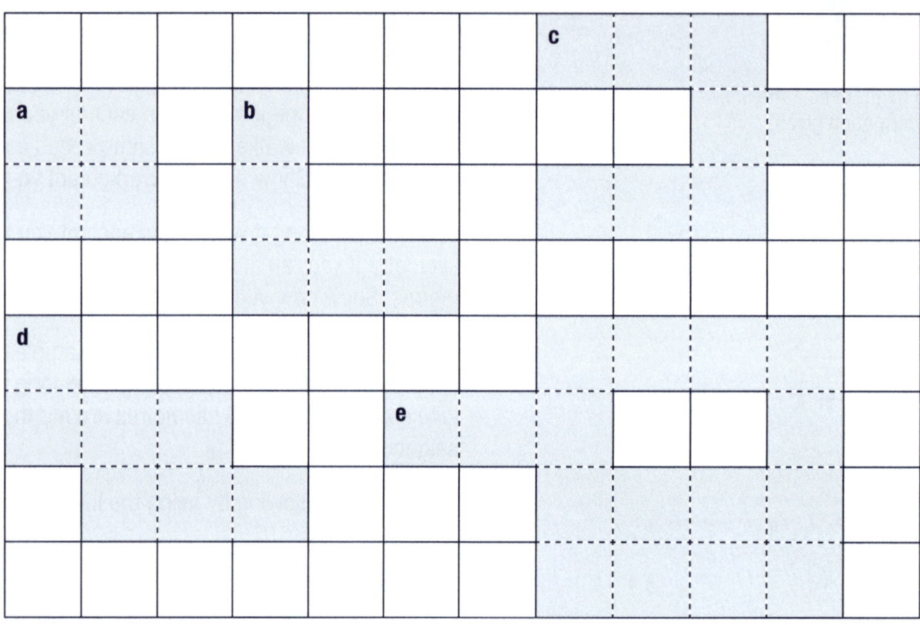

b Calculate the external perimeter of each of the shapes above.

2 Trace the outline of your hand and your foot onto a sheet of grid paper.

a Estimate the two areas then find the approximate areas by counting the squares.

b Write a sentence in your book comparing the two areas.

c Discuss your figures with two other students and decide if there is any relationship between the area of a person's hands and their feet. Write a sentence in your book to describe what you decided and why.

d List other areas that you estimate to be about the same area as your hand. Check to see how close you were.

3 The area of skin on a human body is about 100 times the area of the foot.

a Use this fact to calculate the approximate area of your skin.

b Draw a rectangle and label its dimensions to show what size a rectangle would be that was about equal in area to your body skin.

4 Try to sketch a shape that has an area of 20 cm^2. Check to see how close you were. Write about the method you used to find the actual area of your sketched shape.

Help Box

An easy method for finding the area of a rectangle is to multiply the length of the rectangle by its width.

Area = Length × Width.

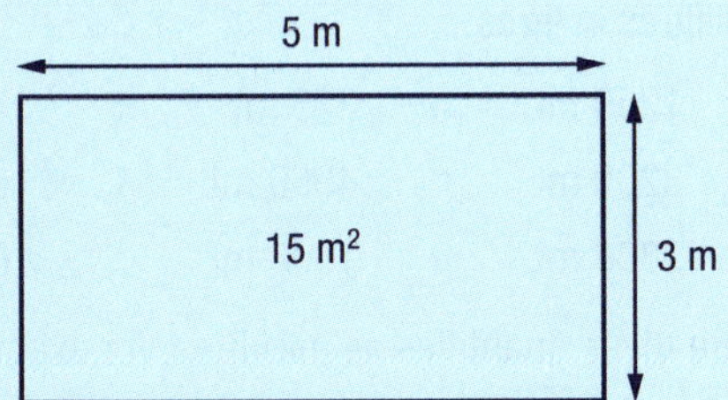

So if a rectangle is 5 m long and 3 m wide

its area can be found by multiplying 5 and 3.

The area is 15 m^2.

5 List some areas that you estimate to be about the same area as the rectangle that you drew in question 3b. Use the method above to see how close you were.

6 Copy this table into your book and complete.

Length	Width	Area
7 m	8 m	
12 cm	5 cm	
8 m		32 m^2
	12 m	48 m^2
10 mm	14 mm	
		24 m^2

7 Use your ruler and draw three different shapes with an area of 24 cm^2. Label the dimensions of the shapes.

8 Use your ruler and draw three shapes that each have the same perimeter but cover a different area. Label the dimensions of the shapes and calculate the perimeter and area of each.

Compound shapes can be divided into two or more rectangles. The area of a compound shape can be found by finding the area of the smaller rectangles and adding them together.

9 Find the area of the smaller rectangles in the following shapes and calculate the total shaded area of each compound shape.

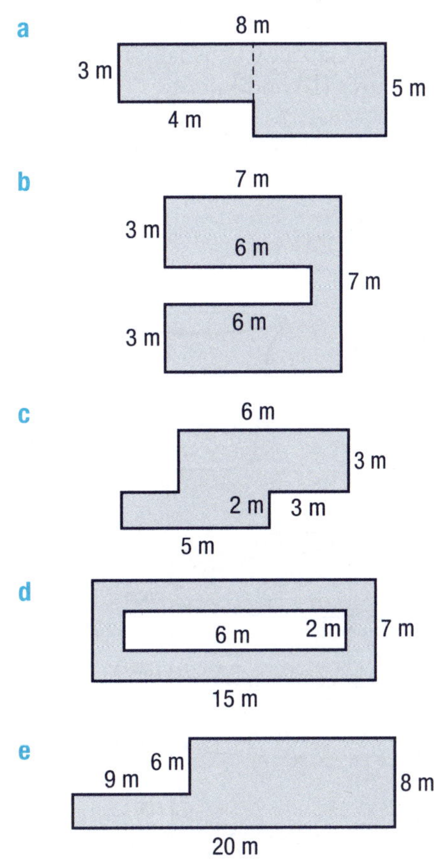

10 Write the name of an area in your school that you estimate to be about the same area as one of the shapes in question 9. Calculate its area to see if your estimate was close.

Lesson 7 Capacity

Athletes know the importance of drinking water to prevent dehydration. About 84% of the human body is made up of water. (Some of this is in the form of blood). People need to drink about two litres of water each day. Capacity is the amount of liquid a container will hold when full. Capacity is usually measured in millilitres (mL), litres (L) and kilolitres (kL).

1 What is the capacity of each of these containers?

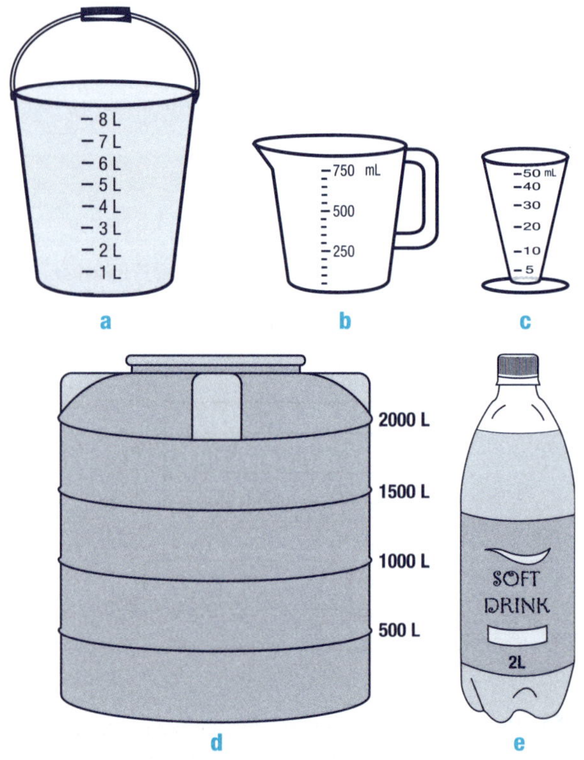

2 How many:

a	mL in 1 L?	b	L in 1 kL?
c	mL in $1\frac{3}{4}$ L?	d	mL in $\frac{1}{4}$ L?
e	L in $\frac{2}{5}$ kL?	f	L in $\frac{4}{5}$ kL?

3 Copy this table into your book and fill in the spaces.

mL	500			2500			
L		1	1.5		3	3.5	5

4 Millilitres can be written as a decimal fraction of a litre (for example: 3500 mL = 3.5 L). Write the following millilitres in litres.

a	1250 mL	b	375 mL	c	1800 mL
d	3200 mL	e	4950 mL	f	1080 mL
g	2866 mL	h	1008 mL	i	1005 mL

5 Write these quantities as millilitres, for example: 3.75 L = 3750 mL.

a	2.5 L	b	8.25 L	c	0.3 L
d	1.356 L	e	4.05 L	f	2.55 L
g	12.05 L	h	0.65 L	i	10.01 L

6 a Estimate then measure how much liquid a cup holds.

b Use a large empty container and water to estimate the following quantities:
250 mL, 400 mL, 750 mL, 1 L.
Check each estimate by pouring the water into a graduated container.

7 Kila kept a record of the amount of fluid that she drank for three days.

Day	Fluid intake
1	400 mL water, 2 cups of tea, 375 mL juice, 750 mL cordial, 250 mL milk
2	3 cups of tea, 700 mL water, 400 mL milk, 250 mL cordial
3	300 mL lemonade, 500 mL water, 2 cups of tea, 500 mL cordial

a Calculate how much fluid Kila drank each day.

b On which day did Kila drink the most fluid?

c What is Kila's average daily fluid intake?

d How does Kila's daily fluid intake compare to the 2 L minimum daily water intake recommended for good health?

8 a List the fluids you drank in the past 24 hours.

b Estimate how much of each fluid you drank.

c Estimate your total fluid intake in the past 24 hours.

d Describe how your fluid intake compares to the average intake you calculated for Kila in the previous question.

9 a Collect data from at least five other students in your class and make a table to show their estimated daily fluid intake and your estimated daily intake.

b Write a few sentences describing some of the differences between the fluid intake of the students in the table and about whether the students drink enough fluid during the day.

10 David and his two friends are doing a fitness circuit. They have 2.5 L of water with them and some mugs that hold 400 mL each.

a How many full mugs of water can they pour from the bottle?

b How many mugs of water will each person get if they share equally?

c How many mL will be left in the bottle?

d How many mugs of water could they each have if the mugs had a capacity of 235 mL instead of 400 mL?

Lesson 8 Relating capacity and volume

Help Box

An empty container with a volume of 1 cm^3 can hold 1 mL of liquid.

Empty containers with a volume of 1000 cm^3 can hold 1 L of liquid.

So, 1000 cm^3 = 1000 mL.

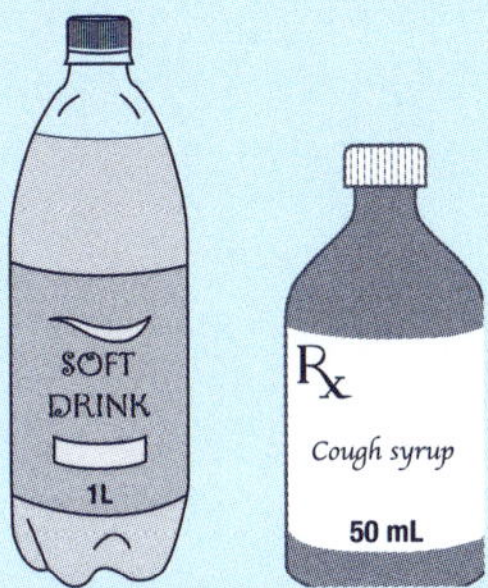

This means that if a medicine bottle holds 50 mL it has a volume of 50 cm^3.

A 1 L bottle of cola has a volume of 1000 cm^3.

Volume measures the amount of space that a three-dimensional (3-D) shape or solid takes up.

Capacity is the amount of liquid a container will hold. There is a link between volume and capacity.

1 What would be the volume of containers with the following capacity?

a 50 mL b 500 mL

c 2 L d 2.5 L

2 What would be the capacity of containers with the following volume?

a 40 cm^3 b 75 cm^3

c 1500 cm^3 d 10 000 cm^3

In the previous lesson you learned that it is important to drink 2 L of water a day. Look back at the table you created in that lesson about how much fluid you and your friends estimate that you each drink per day.

3 a If each student named in the table had to carry their estimated daily fluid intake in one container during a bush trek, what would be the volume of each student's container? Show how you worked out your answer.

b Find a container that is about the same volume as the one required to carry your estimated daily fluid intake.

c Make a table to show how much larger or smaller each student's container would be than the 2000 cm^3 needed to carry the 2 L minimum required for healthy living.

Displacement, volume and capacity

In Grade 6 you explored how to find the volume of regular solids. Displacement is one method that people can use to find the volume of irregular solids.

4 Do the following activity to learn more about the link between volume and capacity.

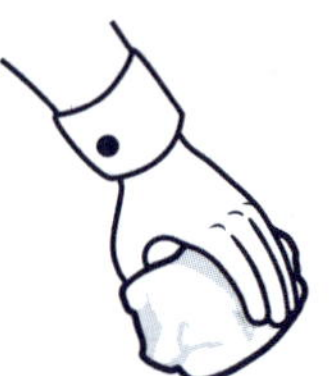

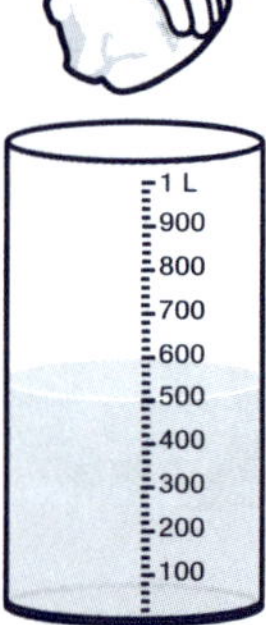

a Collect three stones and rocks of different sizes.

b Estimate the volume of each stone or rock.

c Fill a 1 L measuring container with water to the 500 mL mark.

d Place one stone or rock into the water and write down the new water level on the measurement container.

e The increase in the water level is equal to the volume of the object placed into the container. For example: if the water level rose by 30 mL then the volume of the object is 30 cm^3.

f Repeat the activity to find the volume of the other two rocks or stones.

g How did the actual volumes compare to your estimates?

5 Work with three friends to check who can best estimate the volume of an object. Each find a small rock, estimate your rock's volume and then carry out the displacement activity to find who made the best estimate.

6 a Estimate the amount of liquid you drink while at school in one school day.

b If you had to design a rectangular box to hold your water for the time you were at school on a normal day, what would be the dimensions of the container? Explain your thinking.

Challenge

A megalitre (ML) is a very large unit for measuring capacity.

One megalitre is equal to 1 000 000 litres.

An Olympic-sized swimming pool holds 2.5 ML of water.

- How many litres of water is that?
- What would the volume of the pool be?
- If the bottom doesn't slope what might the pool's dimensions be?

Learning Unit  Revision Challenges

Lesson 1 Revision

1 Write these percentages as fractions in their simplest form.

a 35% b 14% c 25% d 47% e 8% f 10% g 99%

2 Write these fractions as decimals and percentages.

a $\frac{3}{100}$ b $\frac{17}{100}$ c $\frac{8}{10}$ d $\frac{3}{5}$ e $\frac{12}{20}$ f $\frac{2}{7}$ g $\frac{5}{9}$

3 Calculate.

a 3% of K300 b 5% of 4000 L c 10% of 950 g

d 20% of 900 m e 15% of 7500 cm f 25% of K8200

4 A person uses 900 kJ of energy in one hour of walking. How long would the person need to walk in order to use all the energy gained from the following food?

a 120 kJ in a slice of pineapple b 360 kJ from one slice of bread c 1150 kJ from a serve of potato chips

5 Use this food label to answer the following questions:

a How much fat would there be in five slices of this food?

b If I was only allowed 1500 kJ of this food for lunch, how many slices could I eat?

SERVING SIZE (1 slice)	
kJ	315.6 kJ
Protein	2.3 g
Fat	2.7 g
Carbohydrate	15.9 g
Sodium	153 mg

6 These graphs show the different ratios of ingredients two students used to make cordial.

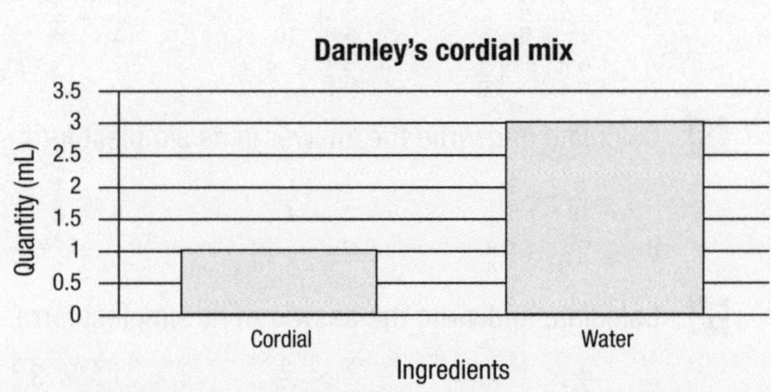

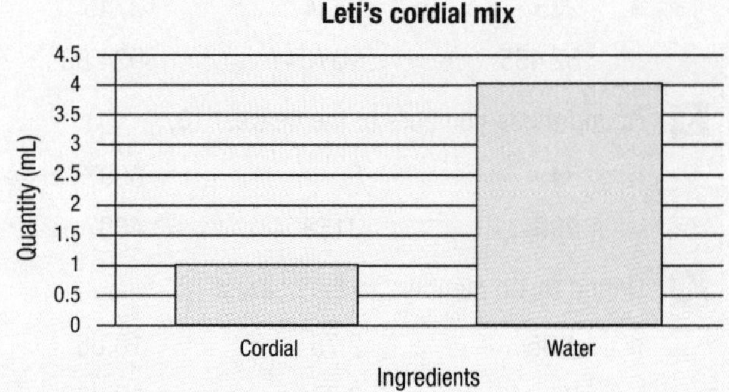

a What is the ratio of cordial to water used by Darnley?

b What is the ratio of cordial to the total mix used by Darnley?

c What is the ratio of cordial to water used by Leti?

d What is the ratio of cordial to the total mix used by Leti?

e What fraction of Darnley's cordial mix is cordial?

f What fraction of Leti's cordial mix is cordial?

g If Darnley was making a container of cordial mix and he was going to use 200 mL of cordial, how much water would he need?

h If Leti was going to make a container of cordial mix and she was going to use 300 mL of cordial, how much water would she need?

i If Leti wanted to fill a 3 L container with cordial mix, how much of the container should be cordial and how much should be water?

7 Write these ratios in their simplest form.

a 3:9 b $\frac{2}{6}$ c 4:12 d $\frac{5}{30}$ e 5:10

f 28:12 g 45:80 h 100:75 i $2:2\frac{1}{2}$

8 A cake shop sells fruit cakes measuring 6 cm by 10 cm. If they sold larger versions of the same cake what would be the dimensions of the larger cake if it increased by the following ratios?

a 1:4 b 2:3 c 2:5

Lesson 2 Revision

1 Draw a number line to show why 3.86 rounds to 3.9 when it is rounded to the nearest tenth.

2 Round these numbers to the nearest 100.

a 235 b 2434 c 3745
d 32 465 e 10 704 f 924.09

3 Round these numbers to the nearest 10.

a 34 b 82 c 6997
d 299 e 3158 f 4097

4 Round these numbers to the nearest $\frac{1}{10}$.

a 4.36 b 2.75 c 18.06
d 35.346 e 2.07 f 10.308

5 Write five numbers that if rounded to the nearest 100 would round to 400.

6 Write five numbers that if rounded to the nearest 10 would round to 80.

7 Write five numbers that if rounded to the nearest $\frac{1}{10}$ would round to 3.2.

8 Estimate answers to these problems and describe the method you used to estimate.

a $267 + 318 + 626$
b $827 - 382$
c 534×23
d $658 \div 33$

9 Calculate the actual answers to question 8. How close were your estimates?

10 Calculate.

a $\frac{2}{3}$ of 36 b $\frac{4}{9}$ of 27 c $\frac{4}{5}$ of 25
d $\frac{3}{4}$ of 52 e $\frac{5}{6}$ of 18

11 If Fabian was $\frac{2}{3}$ of the way through making small cakes for a party and he had baked 14 cakes, how many more cakes did he still need to make?

12 Calculate and write the answer in its simplest form.

a $\frac{2}{3} + \frac{4}{9}$ b $\frac{4}{5} + \frac{3}{10}$ c $\frac{1}{5} + \frac{3}{8}$
d $2\frac{3}{4} + 1\frac{5}{6}$ e $2\frac{4}{7} + 1\frac{7}{10}$

13 Calculate and write the answer in its simplest form.

a $\frac{9}{10} - \frac{3}{5}$ b $\frac{2}{3} - \frac{1}{4}$ c $\frac{7}{8} - \frac{1}{2}$
d $2\frac{1}{6} - 1\frac{3}{4}$ e $2\frac{1}{8} - 1\frac{4}{5}$

14 Calculate and write the answer in its simplest form.

a $\frac{3}{4} \times \frac{1}{8}$ b $\frac{5}{6} \times \frac{9}{10}$ c $1\frac{2}{3} \times 3\frac{5}{8}$
d $\frac{1}{6}$ of $\frac{4}{5}$ e $\frac{2}{7}$ of $2\frac{3}{12}$

15 Calculate and write the answer in its simplest form.

a $\frac{5}{6} \div \frac{1}{8}$ b $\frac{9}{10} \div \frac{1}{2}$ c $\frac{2}{3} \div \frac{3}{4}$
d $2\frac{5}{6} \div \frac{2}{3}$ e $3\frac{5}{8} \div 1\frac{1}{4}$

16 Calculate and convert the answer to a decimal fraction of 1 kg.

a 250 g + 1.25 kg b $1\frac{1}{2}$ kg + 300 g
c 1100 g + $1\frac{3}{4}$ kg d 40 kg − $23\frac{1}{2}$ kg
e $1\frac{1}{4}$ kg − 330 g f $\frac{3}{8}$ kg − 50 g

17 Fabian bought $2\frac{1}{3}$ kg of rice and added it to the $1\frac{7}{8}$ kg of rice that he already had. He then cooked $\frac{3}{8}$ kg of rice. How much uncooked rice does Fabian have left?

18 Annette made five bottles of cordial. Each bottle contained $\frac{3}{4}$ L. She drank $1\frac{5}{8}$ L over the next week. How much cordial does she have left?

Lesson 3 Revision

1 The graphs below have no titles and no labels on the axes. Copy the graphs into your book, then write an appropriate title for each graph and label the axes appropriately. Write three to four sentences about the information on each graph to show how well you are able to interpret the data.

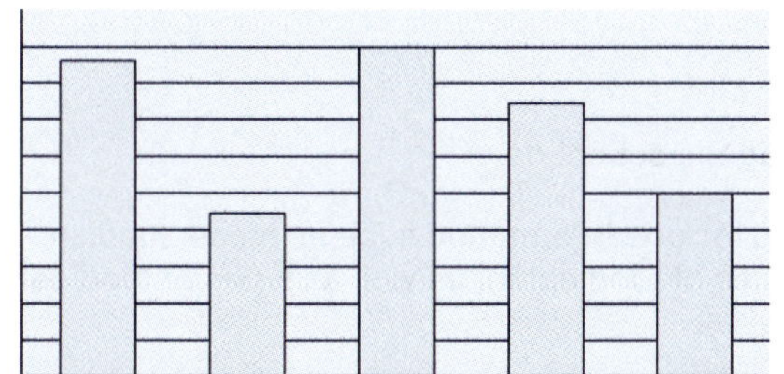

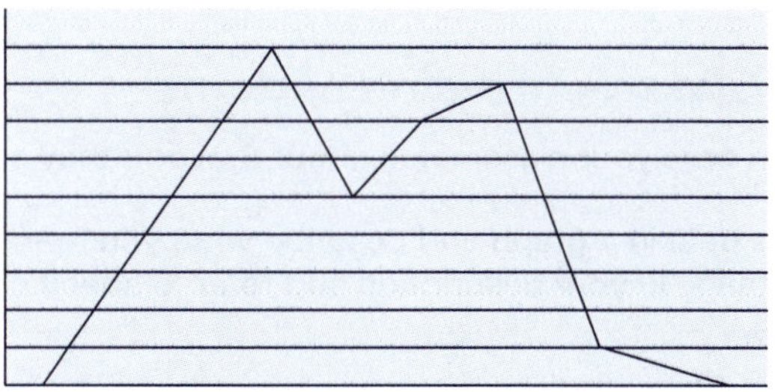

2 Create a pie graph to show what portion of the food that you ate in the past 48 hours came from each of the five food groups; dairy, bread and grains, fruit and vegetables, meat, and fats. Write three to four sentences describing the information in your graph.

3 Find the mean, median, mode and range of the following set of scores:
48, 35, 26, 53, 19, 26

4 A student must select three activities: one from Group A, one from Group B and one from Group C.
Group A activities are basketball and football.
Group B activities are swimming and tennis.
Group C activities are chess and music.
Create a tree diagram to show all the combinations that are available to the student.

5 Select the most likely estimate from the following sets.

- **a** length of a person's natural walking stride: 3 m, 0.7 m, 2.5 m, 140 cm
- **b** weight of an average grade 7 student: 500 g, 5000 g, 50 000 g, 5 000 000 g
- **c** area of a normal school workbook cover: 1000 mm^2, 60 cm^2, 0.4 m^2, 4 m^2.
- **d** capacity of a vase: 3 mL, 300 mL, 0.09 L, 5 L.

6 Copy these statements into your book and complete them.

a 5000 m = ____ km	**b** 1.5 L = ____ mL	**c** 92 kg = ____ g
d 4.2 km = ____ m	**e** 25 g = ____ mg	**f** 25 000 mm = ____ cm
g 4320 mg = ____ g	**h** 0.8 kg = ____ g	**i** 2750 mL = ____ L

7 What is the volume of a container that holds 10 L of liquid?

8 What is the capacity of a container with a volume of 100 cm^3?

9 A walker covered 14 km in 3.5 hours. How far might she:

- **a** have walked in the first 15 minutes if walking at an average pace for the whole 14 km?
- **b** expect to walk in a five hour period at this pace?

Investigations

The first task presented below provides an opportunity to show what you understand about data. The second task provides opportunities to measure weight and length and choose effective strategies. The work you have done in the topic 'Look At Me' will help you to complete these tasks.

Task 1: Classroom eating patterns

Collect data from your classmates about the food they eat during the school day.

Present this data in a graph and describe what you found. Try to include information such as mean, median, mode and range in your description and refer to specific quantities of food eaten, as well as types of food eaten by students.

Make suggestions for how the school might better help students to 'eat healthy' while at school.

If time permits, survey students about how they feel about your suggestions. Include this data in your final report also.

Make a presentation to your class about your work.

Investigations

Task 2: True or false?

Investigate either of the following statements:

I eat the equivalent of my body weight in food every week.

Make a presentation to your class to show what you did.

Include information about the type of food you ate, the quantity and its weight.

OR

Tall people can run faster than short people.

Make a presentation to your class to show what you did.

Include information about the measurements involved and the strategies you used.

Task 3 provides an opportunity for you to show what you understand about ratio and measurement. The work you have done in the topic 'Look At Me' will help you to complete it.

Task 3: Spotlight on my personal fitness/health routine.

Create a table showing an estimation of the main distances that you walked in the last 24 hours. Write an explanation of how you made your estimations.

Create another table showing the food that you ate in the same 24-hour period. Research and/or estimate the number of kJ contained in each item.

Calculate the total kJ you ate for the day and the total km you walked. Prepare a report about your how your kJ intake relates to the amount of walking you did in the day.

Make some suggestions about how you might improve your fitness/health routine in the future. Include a ratio in your description, for example: My exercise rate needs to increase on a 1:2 ratio—I currently walk 2 km a day, but with my kJ intake I may need to walk 4 km a day.

Make a presentation to your class about your work.

Investigations

Task 4 also gives you an opportunity to show your understanding of ratio and measurement. The work you have done in the topic 'Look At Me' will help you to complete this task.

Task 4: What do I now know?

Below are some of the incorrect answers given by Darusila, a grade 7 student, on her test. Read the test questions and Darusila's answers. Work out the correct answers and then write a letter to Darusila to explain where she went wrong and give some advice that might help her get similar questions correct in the future.

1 Ratio

Increase by the ratio 1:3.

a 1 kg beans to 3 kg carrots **b** 30 g flour to 50 mL water

Darusila's answers:

a 1 kg beans, 9 kg carrots **b** 30 g flour, 150 mL water

2 Measurement

Circle the most likely estimate from the following set.

The weight of a newborn baby: 3 g 30 g 300 g 3000 g 0.3 kg

Darusila's answer: 3 g

3 If a person could swim 3 km in 45 minutes, how far would you expect them to swim in 2½ hours if they swim at the same pace for the whole time?

Darusila's answer:

$$3 \text{ km} \times 2\tfrac{1}{2} = (2\tfrac{1}{2} + 2\tfrac{1}{2} + 2\tfrac{1}{2})$$

$$= 7\tfrac{1}{2} \text{ km}$$

Topic 2 Our Country

Learning Unit 1: Which Route?

Learning Unit 2: Water Resources

Learning Unit 3: Below the Ground

Learning Unit 4: Revision Challenges

Topic 2 Our Country

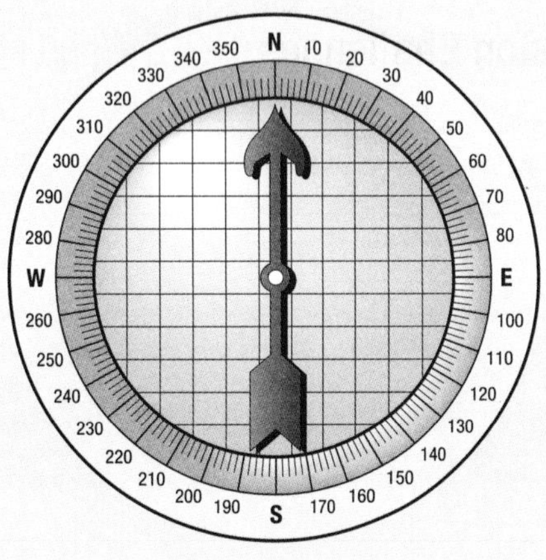

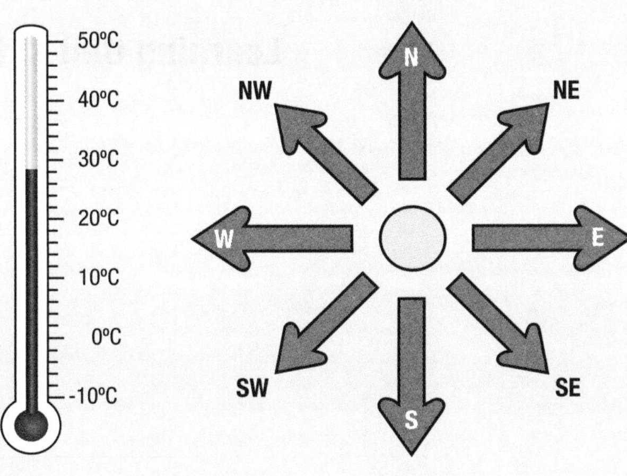

In this topic you will investigate facts about the environment and natural resources of Papua New Guinea.

You will use maps to locate places and to measure distances appropriately. As part of this topic you will also investigate the country's rainfall patterns and the rich mineral deposits and mining industry of Papua New Guinea.

The Revision Challenges provide a range of interesting tasks including the chance to investigate water use within your own home and to extend your thinking about mapping and orientation.

Topic 2: *Our Country* comprises the following Learning Units:

Learning Unit 1: Which Route?

Learning Unit 2: Water Resources

Learning Unit 3: Below the Ground

Learning Unit 4: Revision Challenges

Overview

Each Learning Unit reflects an aspect of mathematics used in everyday life. A brief summary of the mathematics in each Learning Unit appears below.

Learning Unit 1, *Which Route?*, is all about using maps effectively. The unit covers direction and compass bearings (including basic angles), using grids to locate places and calculating distance using the concepts of scale and ratio as they apply to maps.

Learning Unit 2, *Water Resources*, involves working with decimal fractions to make comparisons between rainfall, river flows and water usage in different parts of Papua New Guinea. The unit also explores using roof area to calculate water catchment potential and how this links to the capacity of a water tank.

The work in **Learning Unit 3**, *Below the Ground*, focuses on using numbers below zero to investigate regions below the ground and temperatures below freezing point. This unit also investigates where negative numbers are located on the number line and the number plane. Ratio is revisited when determining the slope of a line on the number plane.

Learning Unit 4, *Revision Challenges*, uses a range of closed short-answer questions, open-ended questions and problems and short investigations to check your understanding of the topic overall.

Some of the materials that you may need to complete the activities in this topic are listed below.

You may need:

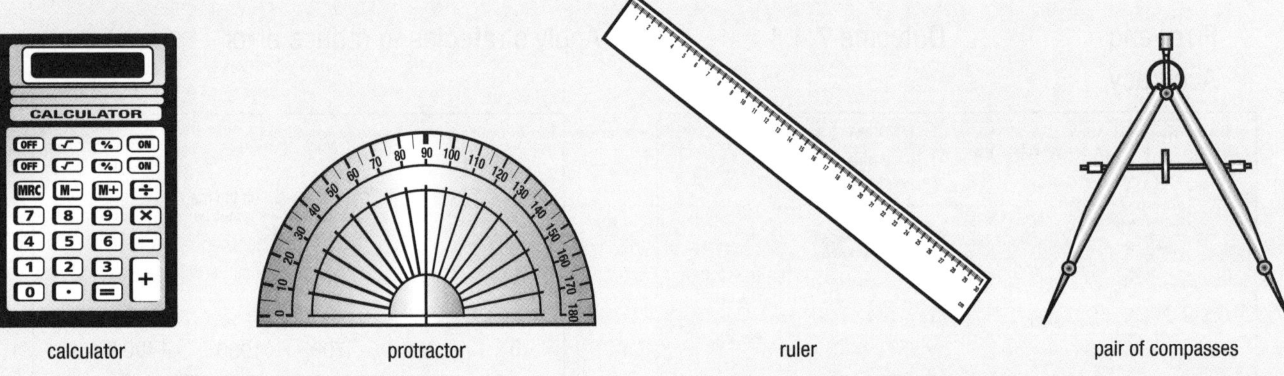

calculator protractor ruler pair of compasses

Learning Unit  1 Which Route?

Strand: Space and Shape

Angles	Outcome 7.2.12	Construct and determine properties of angles
Direction	Outcome 7.2.14	Give the direction of a location relative to others as a bearing
Maps and Coordinates	Outcome 7.2.15	Make maps having scales and keys from suitable data

Strand: Number and Application

Ratios and Rates	Outcome 7.1.5	Convert between ratios and fractions

Strand: Chance and Data

Error and Accuracy	Outcome 7.4.4	Apply strategies to reduce error

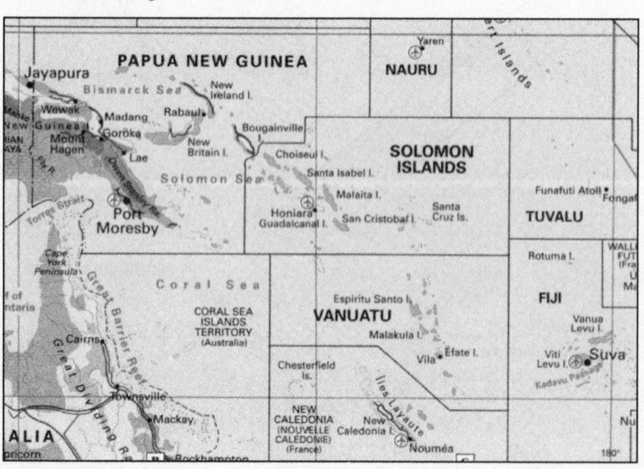

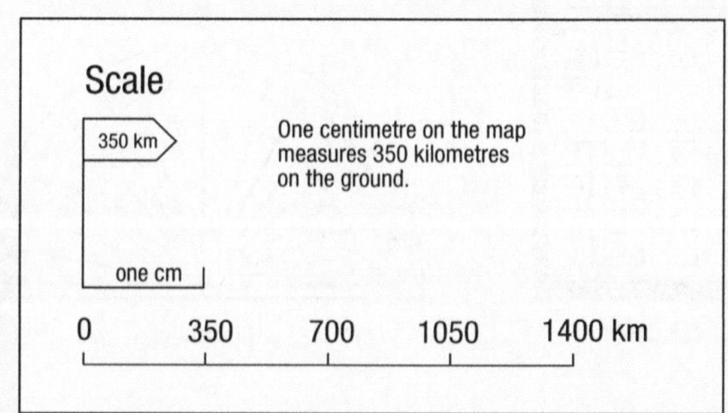

Lesson 1: Introduction	Looking at features of maps
Lesson 2: Coordinates	Locating and giving positions using ordered pairs of numbers
Lesson 3: Direction and turn	Describing direction in different ways Measuring amount of turn with a protractor
Lesson 4: Compass bearings	Using the major and intermediate compass points Using degrees to give bearings Measuring angles
Lesson 5: Naming and constructing angles	Using letters to name angles Measuring and drawing angles with a protractor Using a ruler and a pair of compasses to construct angles Bisecting angles
Lesson 6: Scale	Converting scale measurement to actual distance and vice versa Making maps using suitable scales
Lesson 7: Scale as a ratio	Converting between scale and ratio Using ratio to reduce and enlarge
Lesson 8: Contour lines	Reading contour maps and profiles Drawing profiles from contour maps
Lesson 9: Networks	Drawing simple maps to show how places are connected Deciding if networks are traversable

Lesson 1 Introduction

During this unit you will revise the work you have already done with maps, such as using coordinates and scales. You will use your mathematical knowledge and skills to interpret special features and find locations on maps of Papua New Guinea.

Maps use lines to represent three-dimensional features.

Many maps include:

- a grid that can be used to identify particular coordinates
- a key to help identify particular features
- a scale to help calculate actual distances between locations
- a compass symbol to help identify direction

Use the map of Rabaul to complete the following:

1 Write the name of a street located at the south-east edge of the map.

2 What scale is used on the map?

3 How many kilometres are represented from one side of this map to the other?

4 Using your knowledge of maps, create a key containing at least three symbols that would apply to this map.

5 How could this map be improved?

SCALE 1:13 000

Lesson 2 Coordinates

Maps often include grid lines to more accurately specify a position in two dimensions.

To give a position, we write a pair of coordinates in brackets. The coordinates refer to where the grid lines cross or intersect each other. The order of the coordinates is important. The first number in the bracket refers to the location on the horizontal axis (x-axis) and the second number in the bracket refers to the location on the vertical axis (y-axis).

On this map, the position marked M can be described by the coordinates (6, 3).

The number 6 shows the location on the x-axis and the number 3 shows the location on the y-axis. The location M is at the intersection of the two dotted lines.

The coordinates for the position marked P are (2, 1.5).

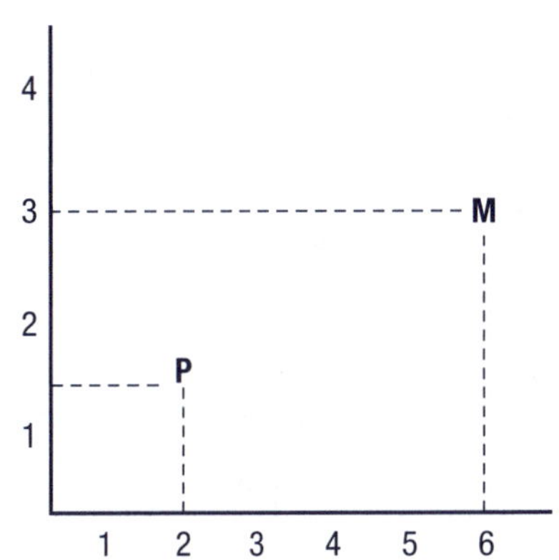

1 Look at the map of Papua New Guinea on the right. Write down the name of the province that has these coordinates:

- a (1, 2)
- b (4, 1)
- c (8, 3)
- d (3.5, 2.5)
- e (4.5, 1.5)

2 List three other provinces and write their coordinates.

3 Write down the coordinates for:

- a Port Moresby
- b Madang
- c Rabaul
- d Wewak
- e Lae

4 Write three sets of coordinates for Western Province (Fly).

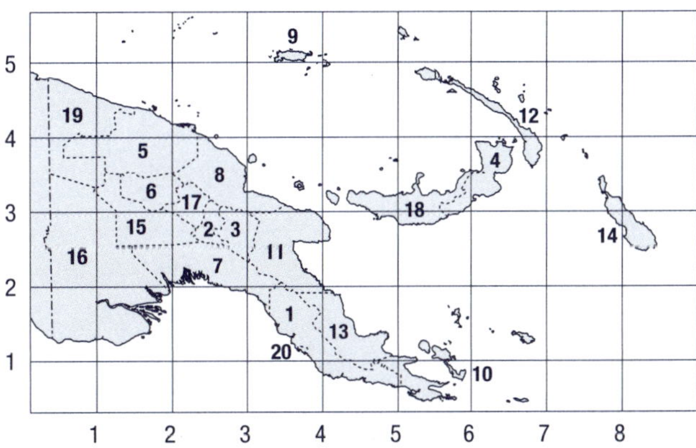

1	Central	11	Morobe
2	Chimbu (Simbu)	12	New Ireland
3	Eastern Highlands	13	Northern (Oro Province)
4	East New Britain	14	Bougainville (North Solomons)
5	East Sepik	15	Southern Highlands
6	Enga	16	Western Province (Fly)
7	Gulf	17	Western Highlands
8	Madang	18	West New Britain
9	Manus	19	West Sepik (Sandaun)
10	Milne Bay	20	National Capital District

5 Look at the grid on the right. Write the letter at these coordinates:

a (6, 5)

b (3, 3)

c (7, 1.5)

d (7.5, 5)

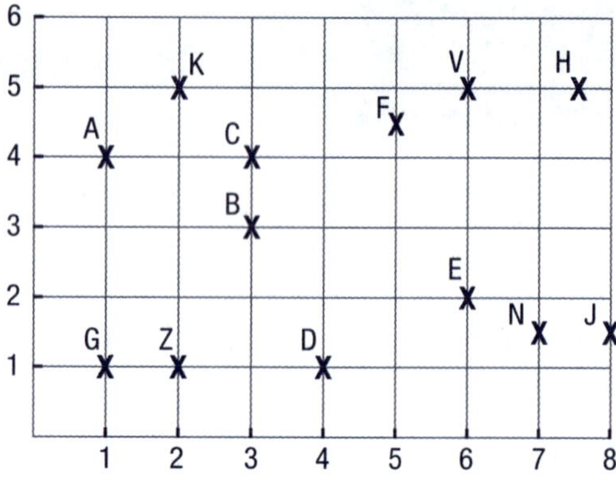

6 Write the coordinates for the other letters on the grid.

7 a Draw a grid like the one in question 5 and locate all the letters of the alphabet on the grid.

b Write the set of coordinates that locate the letters to spell your name.

8 The Botanical Gardens in Lae are located at coordinates (1, 6) on this map. List two other sets of coordinates for the Botanical Gardens.

9 Write coordinates for:

a the Yacht Club

b the Lae Club

c Mt. Lunaman

d the Rugby League Oval

e the Police Station

f the Hospital

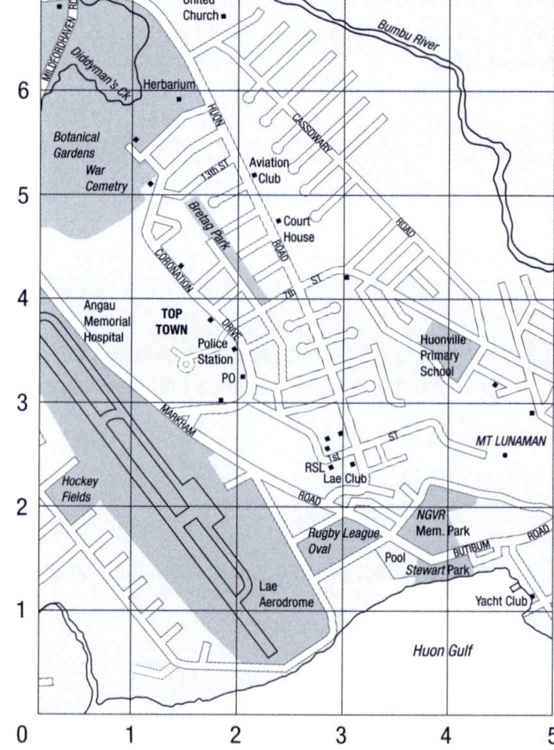

10 Design some questions about the map and give them to a friend to do. Do not forget to work out the answers so that you know if your friend is correct.

11 Draw a map of your classroom (or school) that includes a grid with labelled axes. Write the coordinates to locate five different objects that appear in your map.

Lesson 3 Direction and turn

Often people give directions using their knowledge of fractions and degrees. Phrases such as 'half turn left', 'quarter turn right', '90° turn left' and '180° turn right' are examples of this.

The map on the right shows the location of six children on the beach (A–F). All the children are facing north.

If child A makes a quarter turn to the right she will be facing the boat. If child D makes a half turn to the left he will be facing the shipwreck.

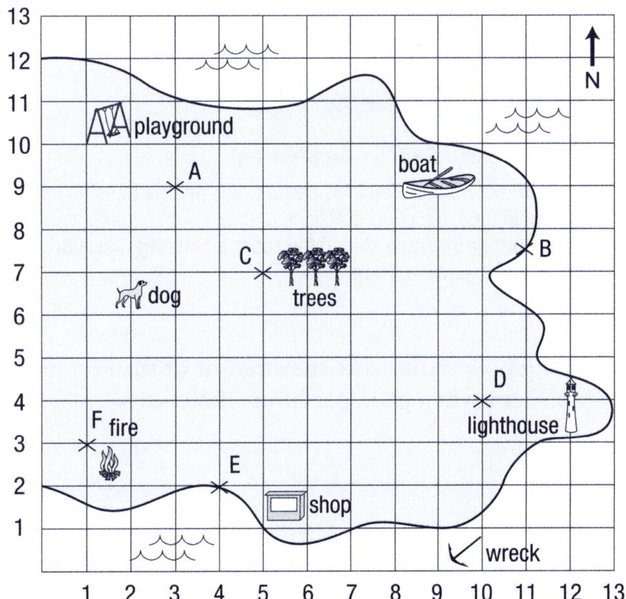

1 Use fractions and/or degrees to describe the movement needed for the following:

- **a** for child B to be facing child C
- **b** for child D to be facing the trees
- **c** for child F to be facing the wreck
- **d** for child C to be facing the shop
- **e** for child A to be facing child C

2 What will each child be facing if he or she makes a half turn to the left?

3 What will each child be facing if he or she makes the following movements?

- **a** child D makes a 90° turn clockwise
- **b** child E makes a 90° turn anticlockwise
- **c** child A makes a $\frac{1}{8}$ turn anticlockwise
- **d** child D makes a 135° turn anticlockwise

4 Bill gave the following instructions to his friend to guide him to draw a shape. Use a ruler and a pencil to mark the following journey:

- Mark a starting point on a piece of paper.
- Go north 5 cm.
- Quarter turn to the right, go east 5 cm.
- Make a 90° turn to the right, go south 5 cm.
- Quarter turn to the right, go west 5 cm.

Which of the shapes shown here did Bill's friend draw?

5 Write instructions using distance, direction, fractions and degrees to help guide a friend to draw one of the other shapes in question 4.

6 List four major towns of Papua New Guinea, other than Port Moresby. Describe the fraction and direction of turn a person standing in Port Moresby, facing north, would need to make in order to face towards each town that you listed.

Challenge

Work with a blindfolded partner. Give instructions using distance, direction, fractions and degrees to guide your partner from one side of the room to the other, avoiding any obstacles. Swap roles.

Help Box

An angle is formed at the point where two lines (called arms) meet. This point is called the vertex.

arm

vertex

arm

Angles can be measured using a protractor. The baseline of the protractor must be lined up with one of the arms of the angle. The centre point of the protractor must be placed on the vertex of the angle.

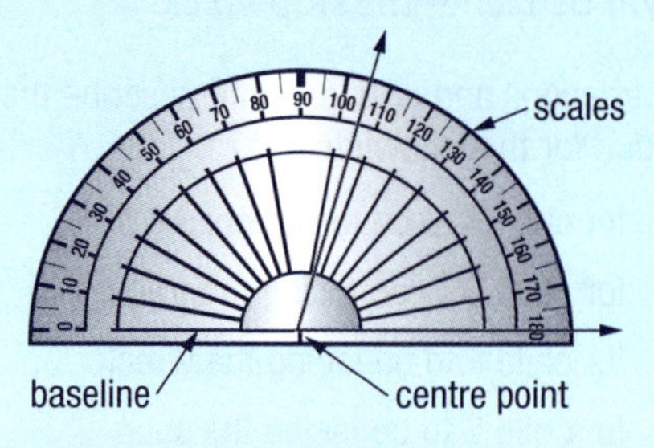

Knowing how to measure the amount of turn (angle) is important when giving accurate directions.

7 Read the protractors below and record the size of each angle in degrees.

a

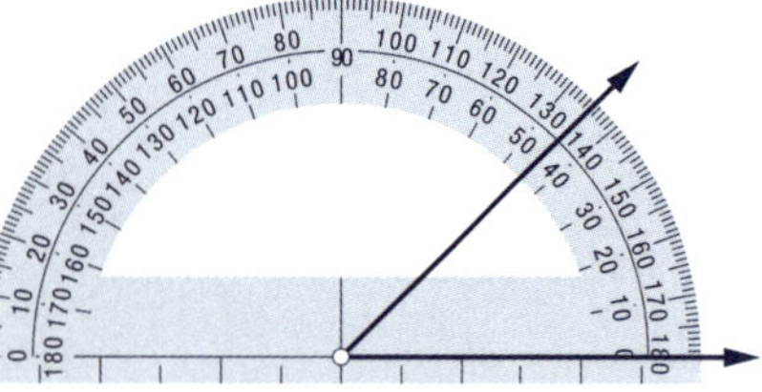

b

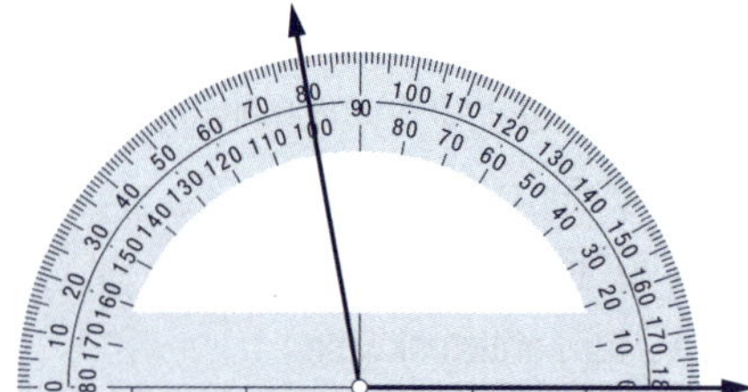

c

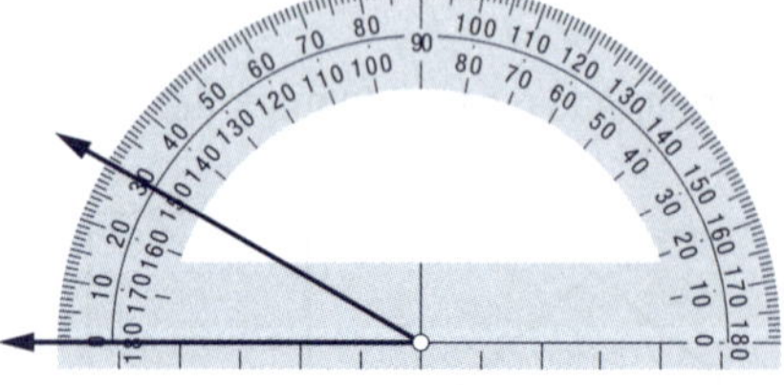

d

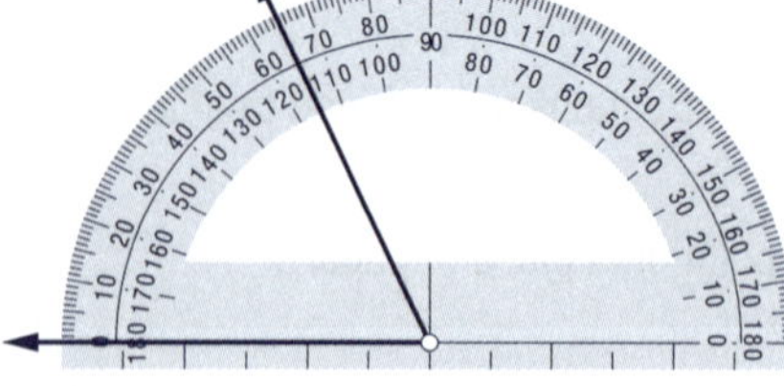

e

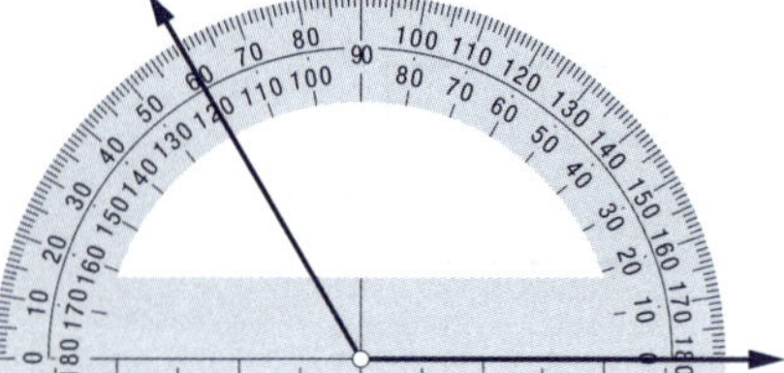

f

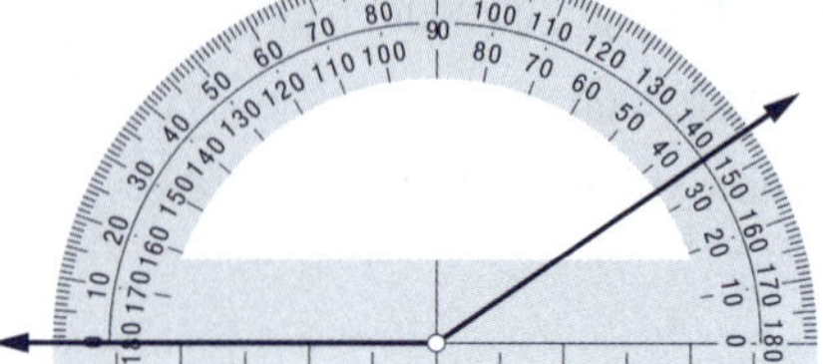

8 Estimate the size of these angles and then measure each one with a protractor to see how close you were.

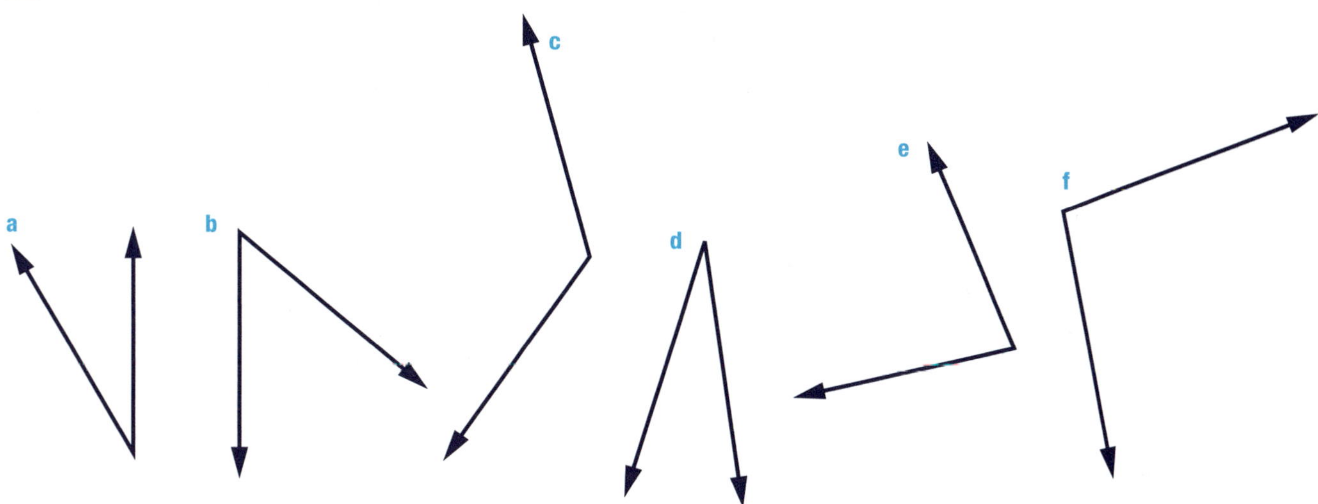

Lesson 4 Compass bearings

Most maps include a symbol showing compass directions (or bearings). Often, only the direction of north is shown. The four main compass bearings are north, south, east and west. The diagram shows four other bearings that exist between the main bearings. North-east (NE), south-east (SE), south-west (SW) and north-west (NW) are called intermediate compass points.

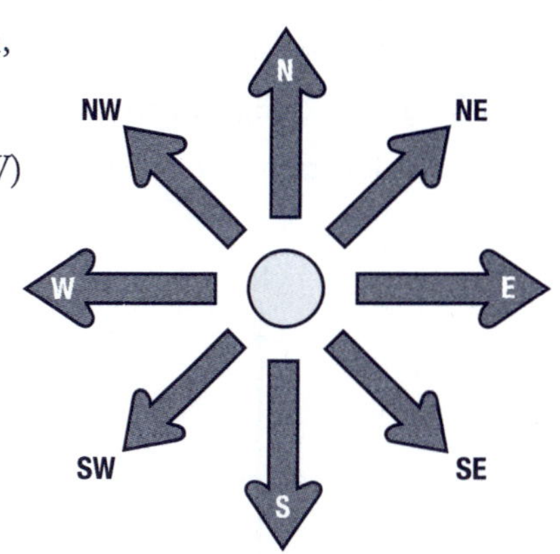

This grid shows different locations marked with a letter.

M	N	O
P	Q	R
S	T	U
V	W	X

1 What letter is:

a south of T? b NE of Q? c SE of P? d north of U?

e SW of N? f NW of W? g north of P? h SE of T?

2 Select one letter on the grid and write three different compass bearings that describe its location in relation to other letters in the grid.

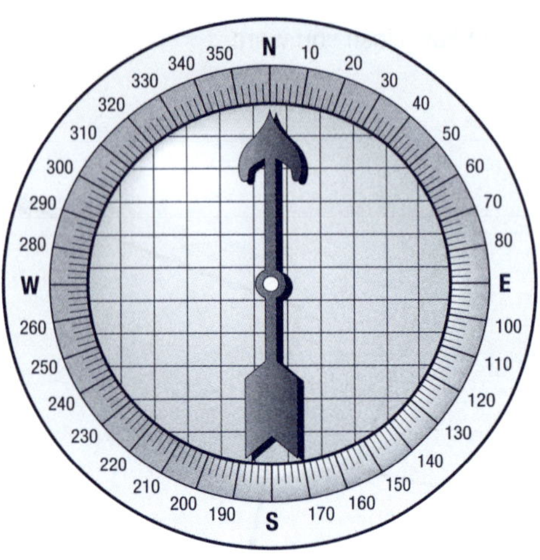

As you know, direction can also be given in degrees. The points on a compass form a perfect circle and can be divided into 360 equal parts.

Measuring by degrees from north in a clockwise direction:

- north is 0° and 360°,
- east is 90° (a right angle),
- south is 180° (a straight angle or two right angles),
- west is 270° (or three right angles),
- and then the 360° is made by returning to the original location (which forms four right angles and is a full circle).

3 Knowing that north is 0°, how many degrees clockwise from north are the following points?

a NE   b SE   c SW   d NW

Direction is usually stated as the number of degrees east or west of the north-south line.

Three things need to be taken into account when writing the direction:

- Will the location be measured in relation to north or south?
- What is the angle of the location in relation to the north-south line?
- Is the location east or west of the north-south line?

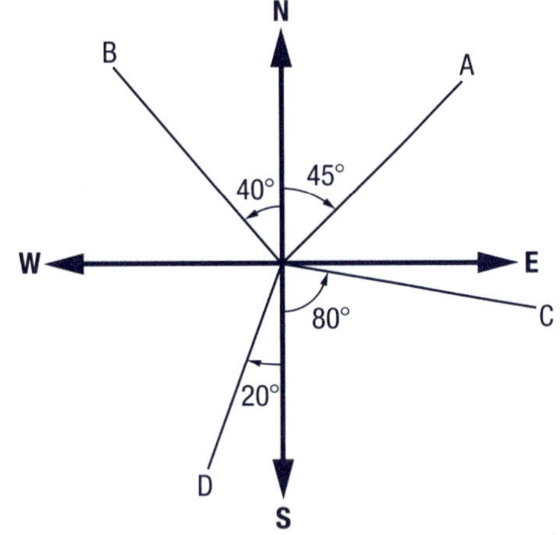

Example: In the diagram on the left, the direction of:

A from the N–S line is N45°E. (N45°E means the direction is 45° east of north.)

B from the N–S line is N40°W. (N40°W means the direction is 40° west of north.)

C from the N–S line is S80°E. (S80°E means the direction is 80° east of south.)

D from the N–S line is S20°W. (S20°W means the direction is 20° west of south.)

4 Use the information in the following diagram to write the direction of the letters K, L, M, O from the N–S line.

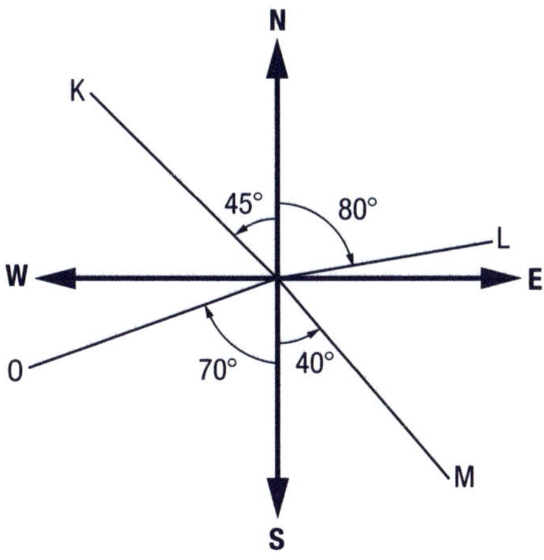

5 Use a protractor to measure the angles on the following diagram and state the direction of the letters M, P, Q and R from the N–S line.

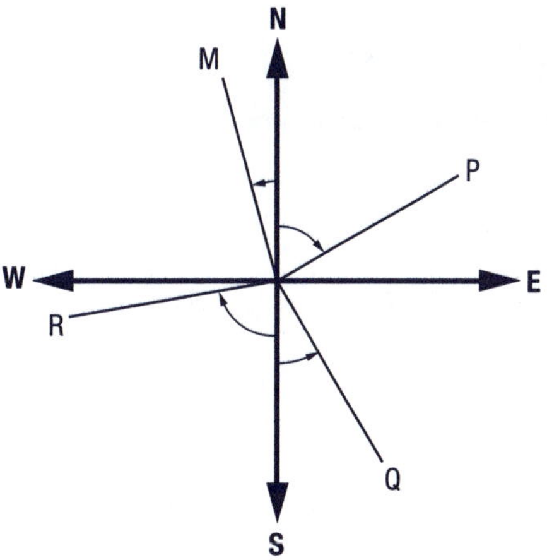

6 Using this map of Papua New Guinea give the bearing of:

a Rabaul from Port Moresby

b Mount Hagen from Lae

c Madang from Port Moresby

d Alotau from Madang

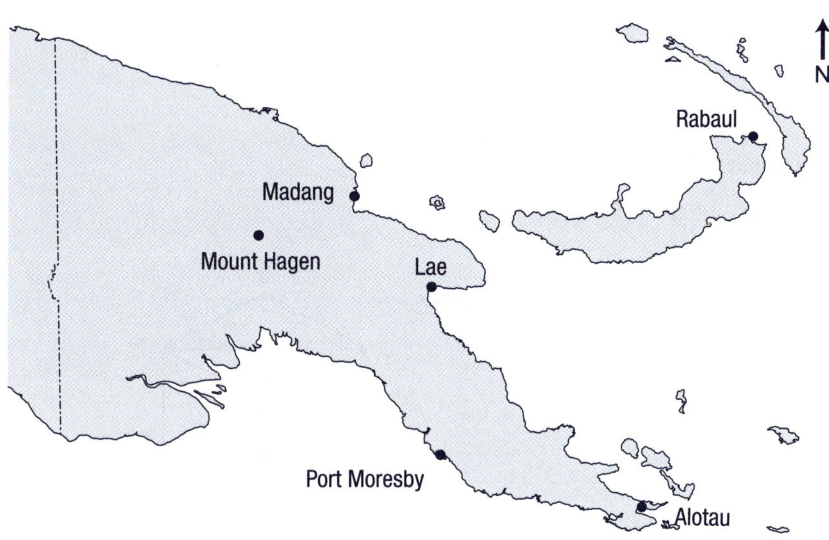

Lesson 5 Naming and constructing angles

Help Box

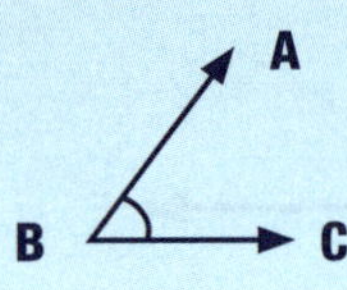

Angles can be named using letters. The symbol for angle is ∠.

The angle shown here can be named ∠ ABC or ∠ CBA.

Notice that the middle letter is always the letter at the vertex of the angle.

The angle could also be known as ∠ B.

1 Write two names that could be used for each of the following angles.

a

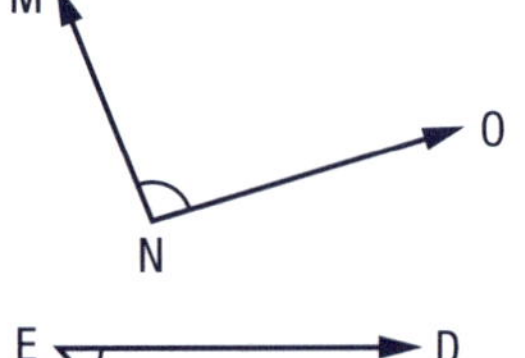

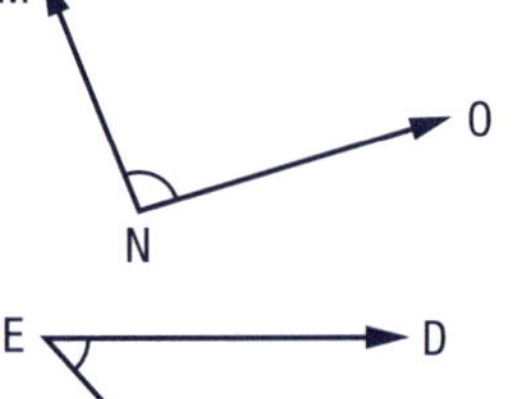

b

c

d

e

f

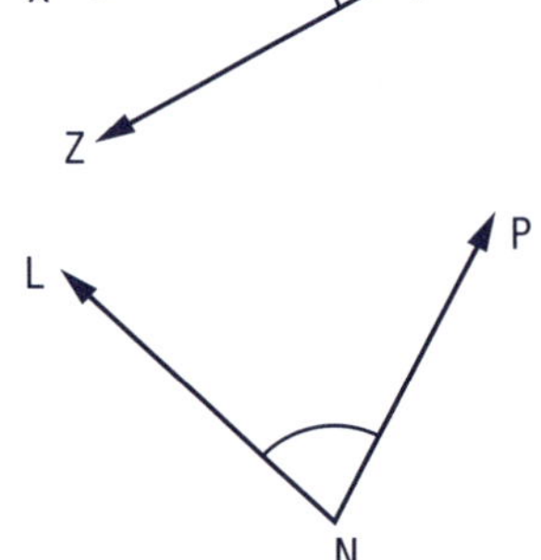

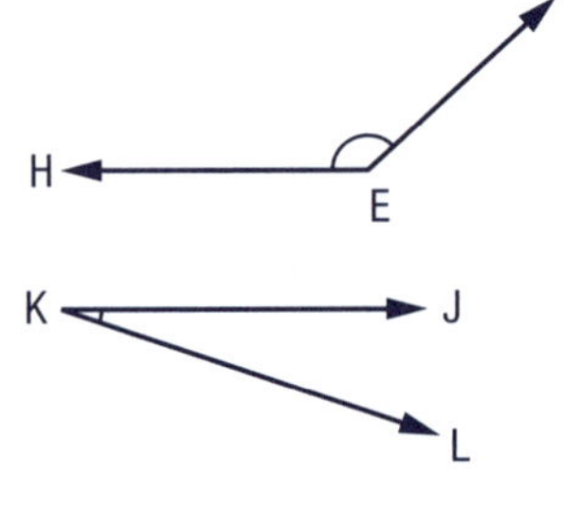

2 Draw a diagram to show each of these angles:

a ∠ ABC **b** ∠ Q

3 Copy this diagram into your book and name each angle that is marked with a symbol.

4 Use a protractor to measure some of the angles in the diagram.

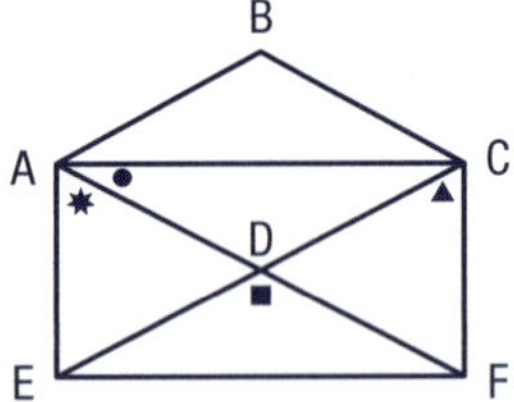

Protractors can be used when drawing maps to ensure features are located accurately. Use a ruler and a protractor and copy the steps below to draw an angle of 45°.

Step 1: Draw a base line and mark one end as the vertex.

Step 2: Place your protractor on the line you have drawn with its centre point on the vertex.

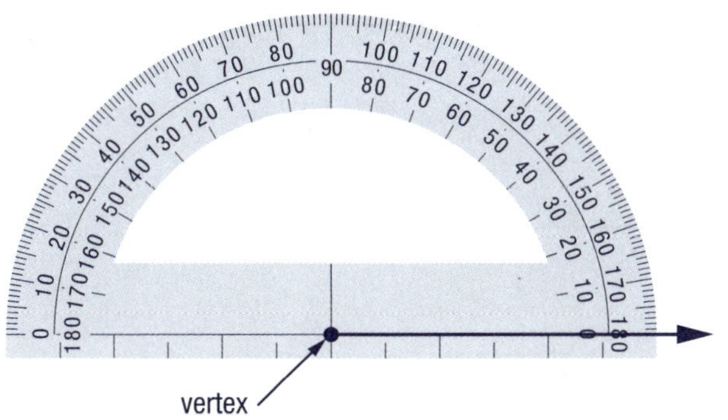

Step 3: Place a mark next to 45° on the protractor.

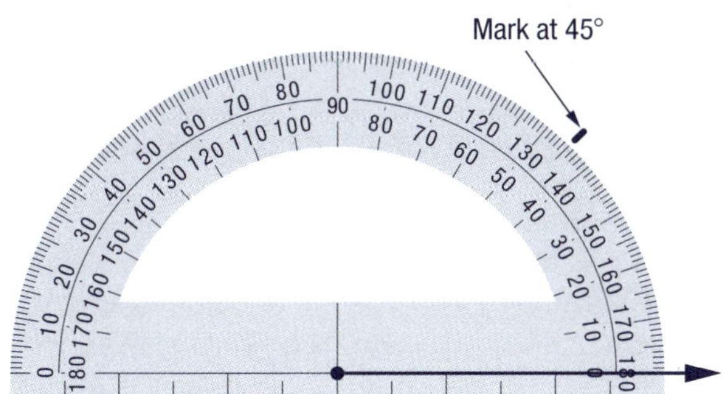

Step 4: Remove the protractor and rule a line from the vertex to the mark.

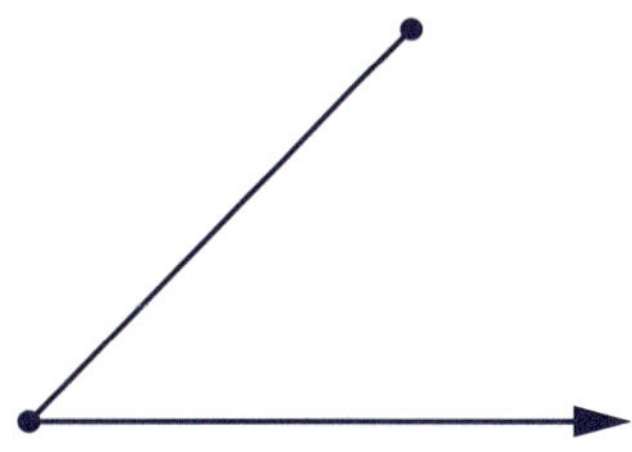

Step 5: Draw an arc to show the angle and write 45° next to it.

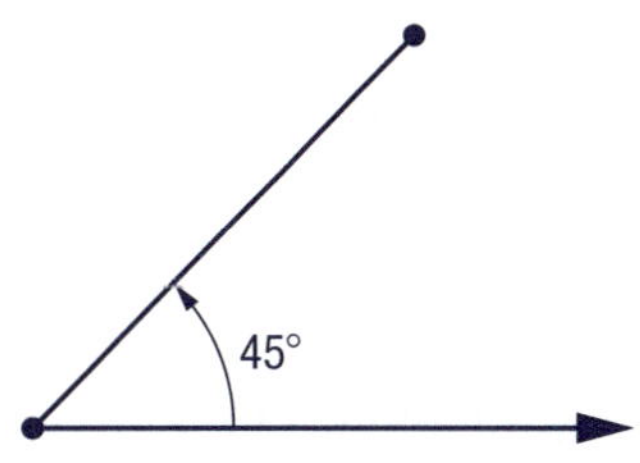

5 Use a ruler and a protractor to draw the following angles.

a 30° b 120° c 90° d 135° e 65°

6 Write each of the angles in question 5 as a compass bearing in relation to the north-south line.

7 Use a ruler and a protractor to draw the angles described in the following compass bearings.

a N45°E b S30°E c N80°W d S10°W e N50°W

8 Find the finishing direction after each turn. The first one is done.

	Start facing	Turn	Degrees	Finish
a	N	clockwise	180°	S
b	S	clockwise	135°	
c	E	anticlockwise		N
d	W	right	45°	
e	NE	left		SW

It is also possible to draw some angles accurately using only a pair of compasses and a ruler. Follow the steps below to draw a 90° angle and a 180° angle using only these tools.

Step 1: Rule a base line 8 cm long and label it AB.

Step 2: Put the compass point on A and open the compass until the pencil point is closer to B than A. Make arcs above and below the line.

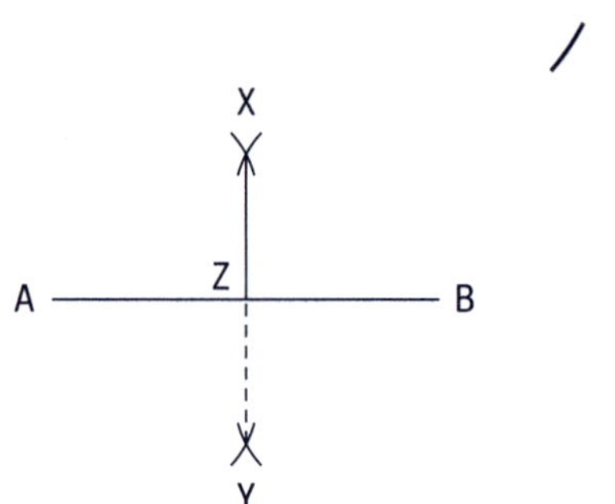

Step 3: Using the same compass setting put the metal point on B and make arcs that intersect the first arcs.

Label the intersection of the arcs X and Y as shown in the diagram.

Rule a line from X to Y.

Label the intersection of line XY with line AB as Z.

$\angle$XZB and $\angle$XZA are 90° and $\angle$AZB is 180°.

9 Name two other angles in the above diagram that are 90°.

It is possible to make a 45° angle by bisecting a 90° angle. Follow the steps below to see how.

Step 1: Begin with the $\angle$XZB that you drew in the previous task.

Step 2: Put the metal point of the compass on Z and draw an arc that intersects the arms ZX and ZB.

Label the intersection points S and T as shown in the diagram.

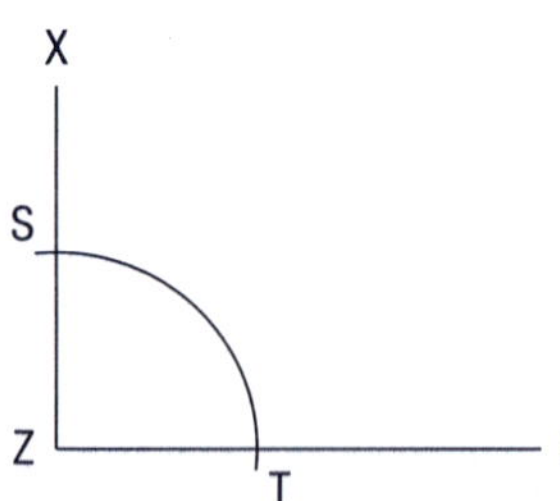

Step 3: Place the metal point of the compass on S and draw an arc between the arms ZX and ZB.

Step 4: Using the same setting on the compass, put the metal point on T and draw an arc to intersect the one drawn in the previous step. Label the intersection point U as shown in the diagram.

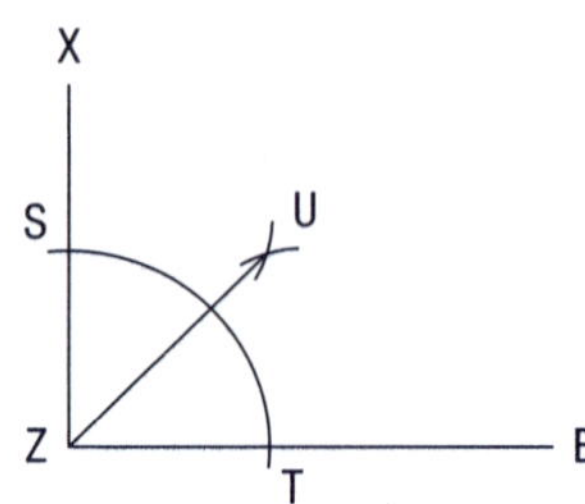

Step 5: Rule a line to connect U and Z. This line bisects $\angle$XZB and forms two equal angles.

10 Name the two angles that are formed.

11 What size is each of these new angles?

To draw 30° and 60° angles follow these steps:

Step 1: Draw a 90° angle and label it FGH.

Step 2: Put the metal point of the compass on G and draw an arc that intersects the arms GF and GH. Label the intersection points M and L as shown in the diagram.

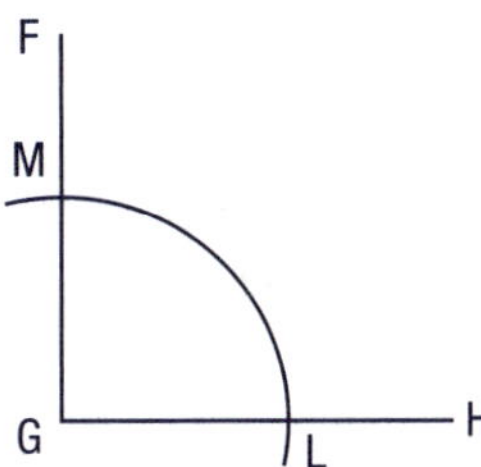

Step 3: Using the same setting on the compass, put the metal point on M and draw an arc to intersect the one drawn in Step 2. Label the intersection P.

Step 4: Still using the same compass setting, put the metal point on L and draw an arc that intersects the one drawn in Step 2. Label the intersection Q.

Step 5: Rule one line to connect G and P and another line to connect G and Q.

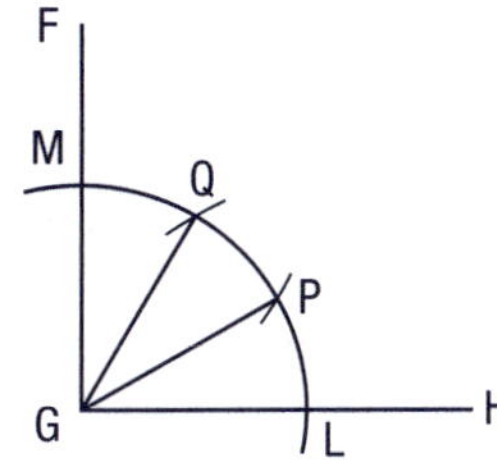

12 What size is ∠ MGQ?

13 Name another angle in the diagram that is the same size as ∠ MGQ.

14 What size is ∠ QGL?

15 What is the sum of ∠ FGQ, ∠ QGP and ∠ PGH?

16 Use your protractor and ruler to draw an angle of 60°. Bisect it using the method you used above to bisect an angle of 90°. What do each of the angles measure? What is the sum of their degrees?

17 Draw an angle larger than 90° with your protractor and ruler. Bisect the angle as described previously. What do each of the new angles measure? What is the sum of their degrees?

Lesson 6 Scale

The scale on a map shows the relationship between the distance on the map and the actual distance in real life. On the map below showing Port Moresby Town (Map A), one centimetre represents 130 metres in real life. On the map of Papua New Guinea (Map B), one centimetre represents 1080 kilometres.

Map A

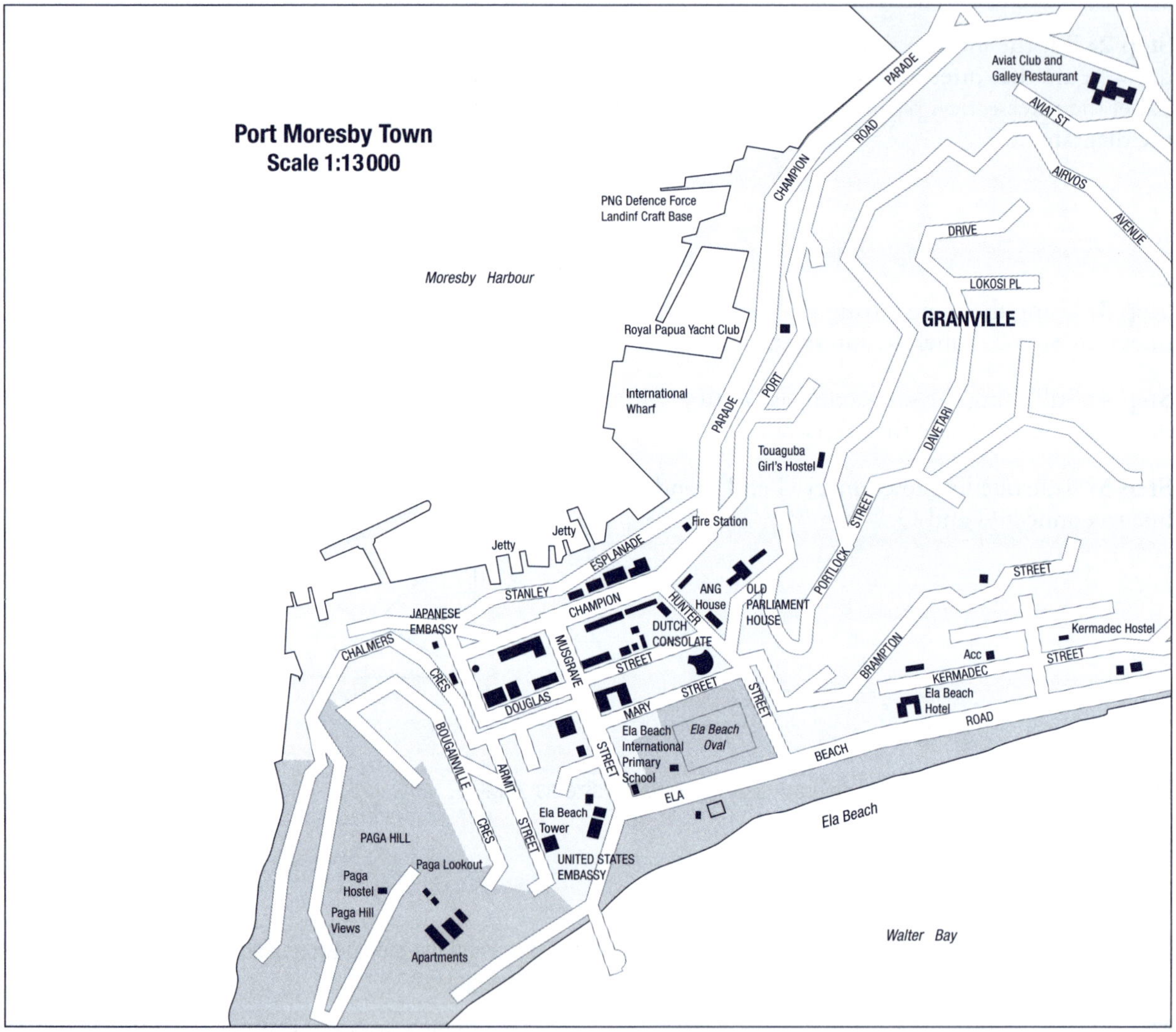

Port Moresby Town
Scale 1:13 000

Map B

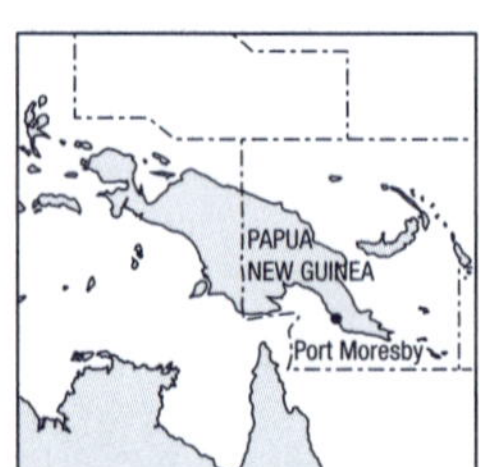

Papua New Guinea
Scale 1:108 000 000

1 Answer the following questions about Map A.

a What length would a road that was actually 13 km long be on the map?

b If two locations are 5 cm apart on the map, how far apart are they in real life?

2 Use your ruler and the scale marked on Map A to find the actual distance from:

a Bougainville Crescent to Hunter Street

b Stanley Esplanade to Ela Beach Road

c one end of Chalmers Crescent to the other (Paga Lookout to Musgrave St.)

3 Name two places on Map A that are:

a about 2 km apart

b about 0.5 km apart

c more than 2 km apart

4 Answer the following questions about Map B.

a What length would a distance of approximately 500 km appear on this map?

b If two locations are 3 cm apart on the map, how far apart are they in real life?

5 Use your ruler and the scale marked on Map B to find the actual distance from:

a Port Moresby to the western border of Papua New Guinea

b Port Moresby to the northern tip of Australia's mainland

6 Using the scale 2 cm = 500 m, calculate the measurements to be used on a map for the following real distances.

a 1 km b $5\frac{1}{2}$ km c $12\frac{1}{4}$ km

d 250 m e 1000 m f 4.5 km

g 5.25 km h 10.25 km i 9.75 km

7 Copy the following tables into your book. Use what you know about scale to complete the information.

a Scale 1 cm = 5 km

cm on map	1	3		0.5	
actual km	5		50		62.5

b Scale 1 cm = 3 km

cm on map	1		7		23.5
actual km	3	15		7.5	

8 Measure the length of the three lines below. Then copy the table into your book and complete it to show what each line would represent on different scaled maps.

a ____________

b ____________________

c ___________________________________

Line	Scale 1 cm = 2 km	Scale 1 cm = 500 m
a		
b		
c		

9 a Make a map of your desk top and select a scale that would be appropriate for the amount of detail to be shown.

b Explain how you decided what scale to use for your map.

10 a Make a map of your classroom and select a scale that would be appropriate for the amount of detail to be shown.

b Explain how you decided what scale to use for your map.

11 If you were making a map of the clan land in a village, what scale would you choose? Why?

Lesson 7 Scale as a ratio

In Topic 1, Learning Unit 1, *Food and Nutrition*, you did some work on ratio with regard to recipes. Ratio is a way of comparing amounts. Scale can also be written as a ratio.

When writing a scale as a ratio it is important to convert both amounts to the same unit of measurement.

Example: A scale on a map is 1 cm = 1 km.

The scale can be written as the ratio 1: 100 000.

This means that every 1 cm on the map represents 100 000 cm (or 1 km) in actual distance.

SCALE 1 cm = 1 km is the same as RATIO 1:100 000

1 Convert the following measurements to the same unit of measurement and rewrite the scale as a ratio.

- a 1 cm = 2 km
- b 1 cm = 10 cm
- c 1 cm = 1 m
- d 1 cm = 500 m
- e 2 cm = 1 km
- f 2 cm = 20 km
- g 2 cm = 6 cm
- h 3 cm = 10 cm
- i 4 cm = 2 m

2 A map is drawn using a scale 1:2000. Calculate the real distances in metres for the following distances measured from the map.

- a 2 cm
- b 5 cm
- c 6 cm
- d 6 mm
- e 18 mm
- f 19 cm
- g 0.5 cm
- h 1.2 cm
- i 4.1 cm

3 The scale on a map is 1: 2 500 000.

- a What does this mean?
- b How far in actual distance would 1 cm, 10 cm and 30 cm on the map be?

Help Box

Maps can be enlarged or reduced by using a ratio.

Example: Enlarging a map on a 3:5 ratio, means that every 3 cm on the original map will become 5 cm on the enlarged map. The original map will be $\frac{3}{5}$ the size of the enlarged map.

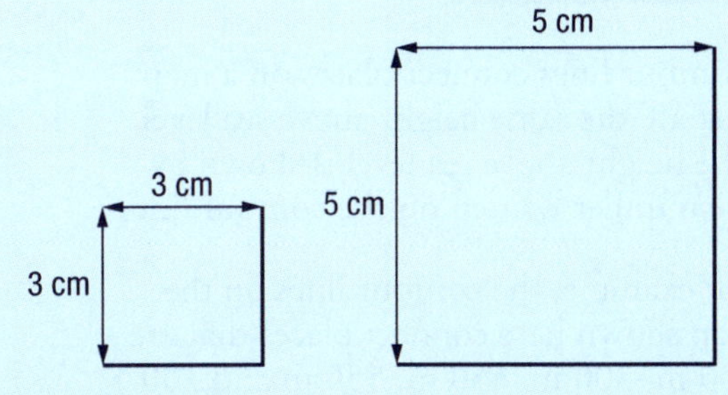

4 A small inset map measuring 2 cm by 3 cm is to be enlarged. Copy the table into your book and complete it to show the dimensions for enlargements with the same ratio.

Type of map	Width	Length	Ratio
Inset size	2	3	2:3
Pocket sized map	8		2:3
Page sized map		21	2:3
Small poster size	24		2:3

5 Measure and then redraw this grid into your workbook using a 1:2 ratio. Use the grid to help you enlarge the map of Papua New Guinea using the same ratio.

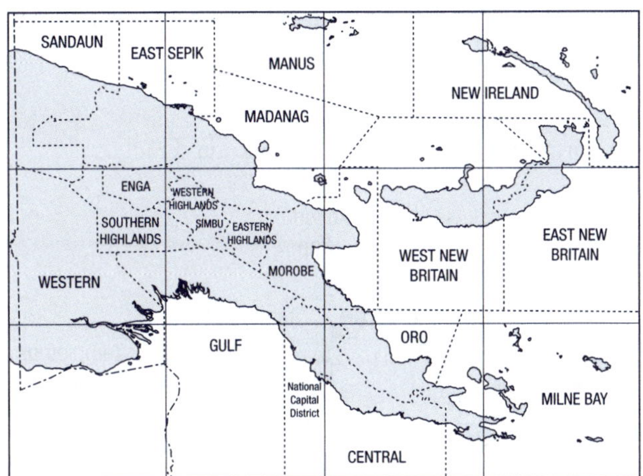

Lesson 8 Contour lines

Contour lines connect places on a map that are the same height above sea level. The height above sea level is shown by the number written on the contour line.

For example, the contour lines on the map shown here connect places that are 200 m, 300 m, 450 m, 550 m and 700 m above sea level.

The town located at point E is 300 m above sea level because it lies on the 300 m contour line. The town located at point A on the map is between 450 m and 550 m above sea level.

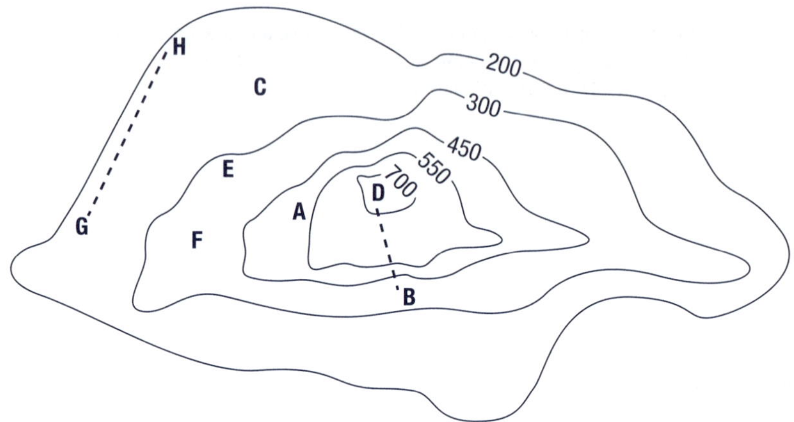

1 Use the map to answer the following questions:

a How high above sea level is town F?

b How high above sea level is town C?

c List two towns that are the same height above sea level.

d Which town on the map is located at the highest point above sea level?

e Describe the difference betweeen walking the path from towns H to G compared to the path from towns B to D in relation to effort required.

Betty drew the sketch to show the profile of one part of the Kokoda Track.

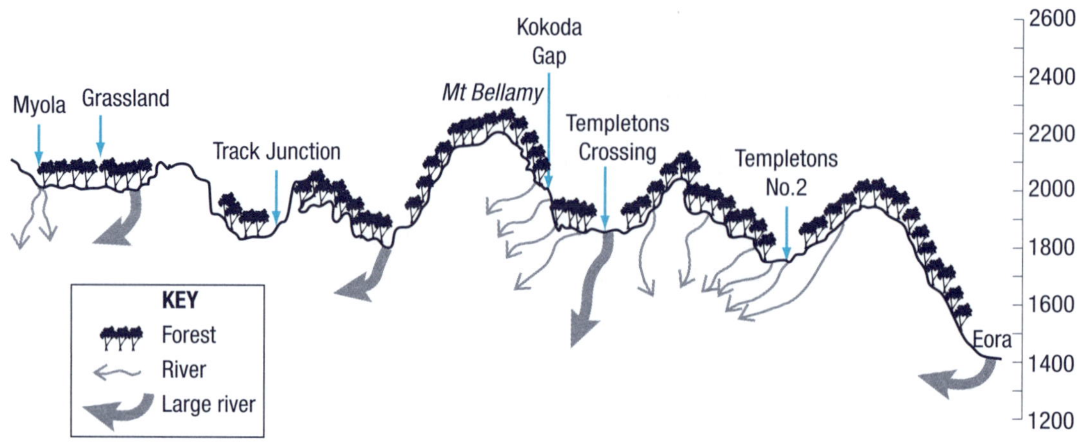

2 Use Betty's sketch to answer the following questions.

a What height above sea level is the Kokoda Gap?

b What is the difference between the height above sea level at Myola and Eora?

c How high is Mt Bellamy?

d What is the difference between the height above sea level at the start and the end of this section of the track?

Here is a contour map that Betty drew to show a small section of the profile she had drawn in the previous sketch.

3 Match the information on the contour map and the profile to answer the following questions.

a What is located at the position marked A on the contour map?

b What is located at position B on the contour map?

c What is located at position C on the contour map?

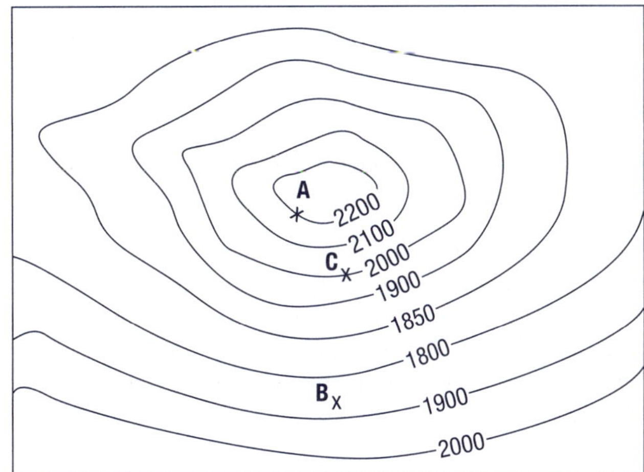

On the right is a contour map of an imaginary island.

4 a Sketch a profile of what the island might look like if you were looking at it from the south.

b Sketch a profile of what the island might look like if you were anchored in a boat off the SE corner of the island.

c Write a few sentences to describe the lay of the land if you were viewing the island from the west.

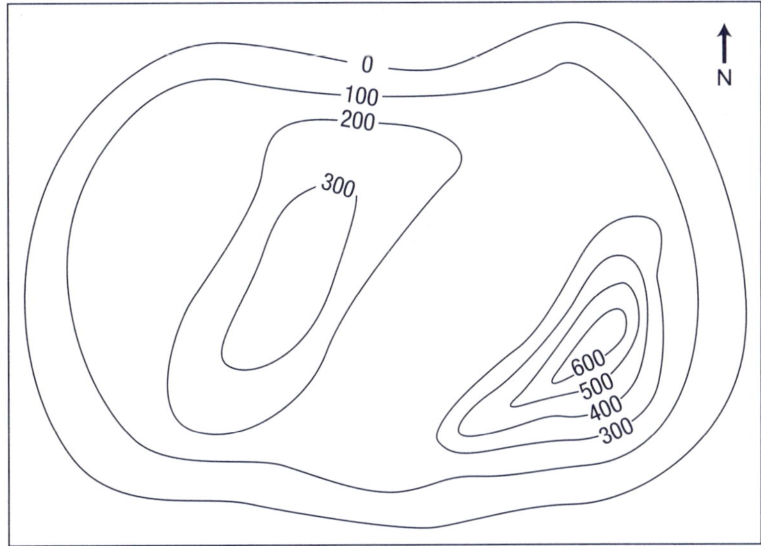

Lesson 9 Networks

Sometimes simple maps are drawn that show the connections between places. These maps often do not have specific measurements included and may not even be drawn to scale. These simple maps are called networks.

This is a sketch of the network that Marlon's father drew to show him two different ways that he might get from their house to the market.

1 Draw a network to show two different routes that you might take to get from your home to school. Include some of the landmarks that you pass during the journey to show how different places are connected.

Below is a network drawing of the tracks connecting five small villages. The villages are marked by dots, called *nodes*, and the connecting paths are shown by lines called *arcs*. The position of the nodes and the length of the arcs are not important on a network drawing.

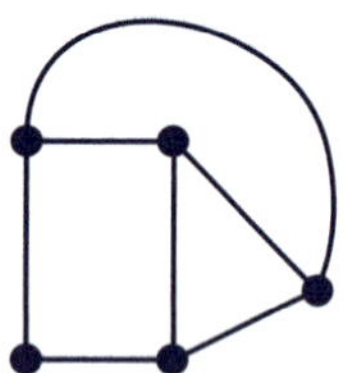

Identical networks might therefore look very different. The following network drawing could also represent the network above.

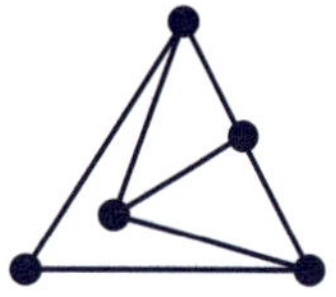

Help Box

Labelling the nodes (A, B, C ...) when drawing networks can make it easier to check the connections between them.

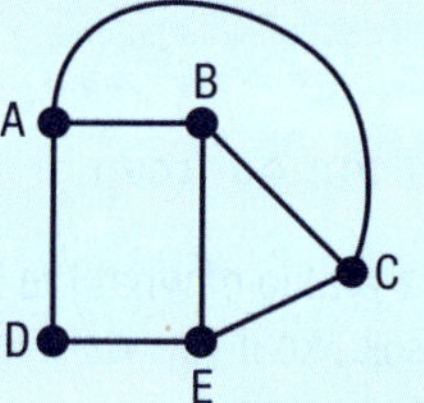

Example:

A connects to BCD

B connects to AEC

C connects to ABE

D connects to AE

E connects to BCD

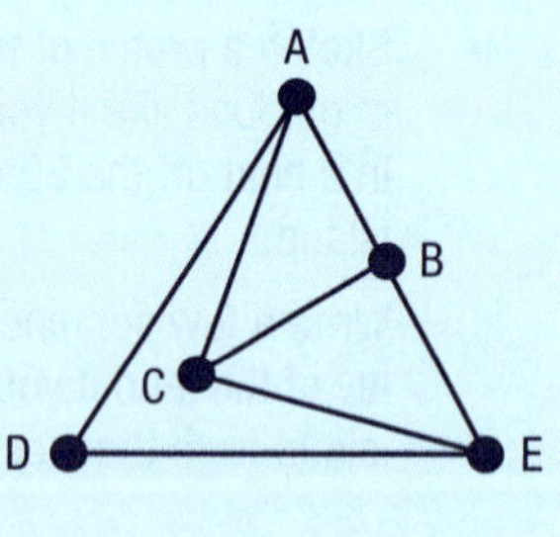

2 Copy each of the following networks into your book. Label the nodes and then redraw each network in a different way.

a
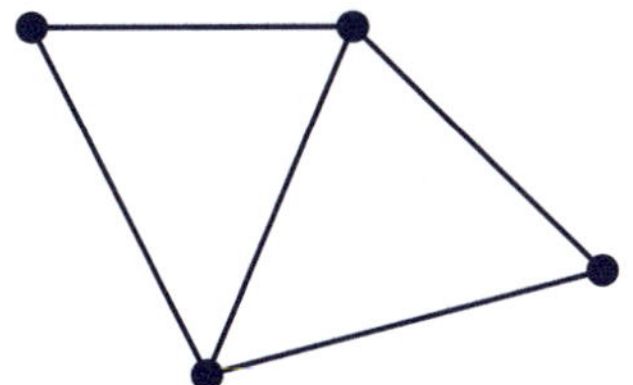
b
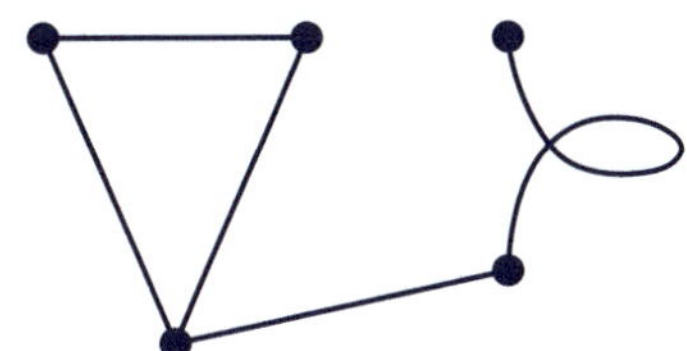
c
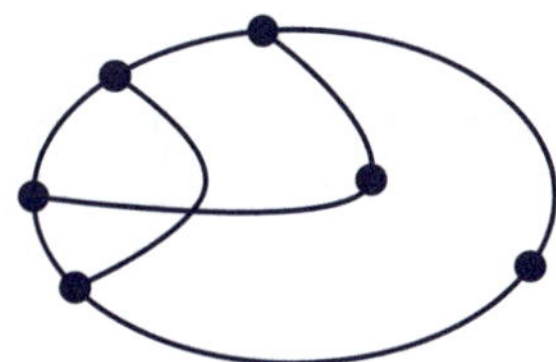

3 Make a table for each network to show that the nodes in your redrawn version match the connections in the original network.

4 Draw two different network maps that show the connections between the towns on the following map. It will be easier if you label the nodes first.

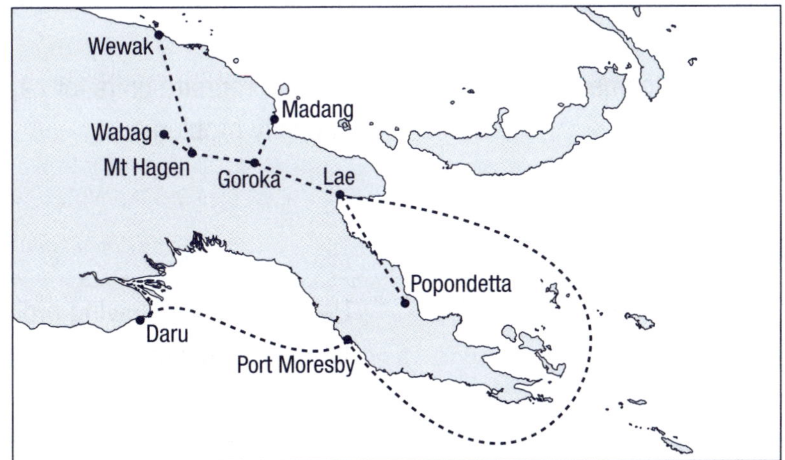

5 Some networks are *traversable*. This means that it is possible to travel along each arc and pass through each node only once. Copy the network maps below and experiment to decide which are traversable. Label the nodes on the networks you draw to make it easier to check the connections.

a

b
c
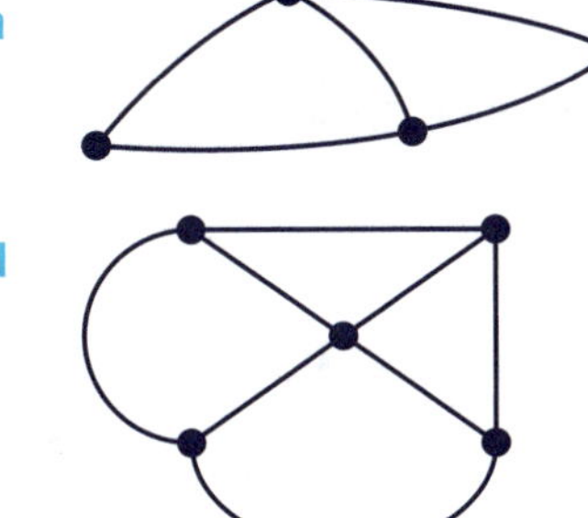
d
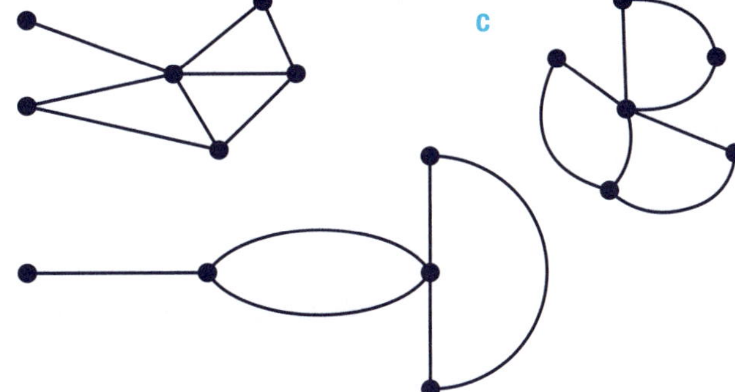
e

6 Draw a network of your own that is traversable.

7 Choose two of the networks in question 5 and redraw them so that the overall shape changes but the network connections remain.

8 Examine the network drawings you completed in question 2 and decide if they are traversable.

Learning Unit  Water Resources

Strand: Space and Shape

Length	Outcome 7.2.2	Use appropriate metric units in calculations
Area	Outcome 7.2.4	Compare areas by estimation
Capacity	Outcome 7.2.8	Use appropriate units for capacity and solve capacity problems

Strand: Number and Application

Decimals	Outcome 7.1.2	Use decimals in solving problems set in familiar contexts
Fractions and Decimals	Outcome 7.1.3	Convert between fractions, decimals and percentages

Strand: Chance and Data

Error and Accuracy	Outcome 7.4.4	Apply strategies to reduce error
Estimation	Outcome 7.4.6	Use a variety of estimation strategies

Lesson 1: Introduction	Reading graphs about water usage and fishing in Papua New Guinea
Lesson 2: Understanding, comparing and ordering decimals	Using place value to represent, compare and order decimals
Lesson 3: Rounding decimals	Rounding to whole numbers Rounding to same number of decimal places
Lesson 4: Adding and subtracting decimals	Writing litres and millilitres as decimals Using addition and subtraction to solve decimal problems
Lesson 5: Multiplying decimals	Multiplying decimals by 10, 100, 1000 Multiplying decimals by multiples of 10 Multiplying decimals by whole numbers other than 10 Multiplying decimals by decimals
Lesson 6: Dividing decimals	Dividing decimals by 10, 100, 1000 Dividing decimals by multiples of 10 Dividing decimals by whole numbers other than 10 Dividing decimals by decimals
Lesson 7: Measuring rainfall	Interpreting rainfall data Reading scales on rain gauges
Lesson 8: Area of roofs and water catchment	Using decimals to estimate and calculate area
Lesson 9: Solving problems	Using whole numbers and decimals to solve problems related to length, area and capacity
Lesson 10: Calculating the amount of water to be caught	Relating rainfall, roof area and catchment capacity

Lesson 1 Introduction

In this unit you will learn about some of the different ways that people in Papua New Guinea use water in their daily lives, and find how valuable water tanks are for storing water. You will also learn how to use decimals to calculate solutions to different problems.

Papua New Guinea's humid climate means there is an abundance of water. However, because the wet season only lasts for around four months each year, seasonal water shortages are common. Most urban towns have clean drinking water but this is not the case for over 70 per cent of the rural population.

From the high mountain peaks, great rivers like the Sepik and Fly Rivers begin their journey to the sea. These rivers support an enormous fish population as well as providing water for the community.

This pie graph shows the different ways that water is used in Papua New Guinea.

1 Write a few sentences to explain what you understand from the information contained in the pie graph.

2 Make a list of the different ways your family uses water each day, and estimate how much water each item uses.

3 Draw a pie graph to display the information about your family's use of water.

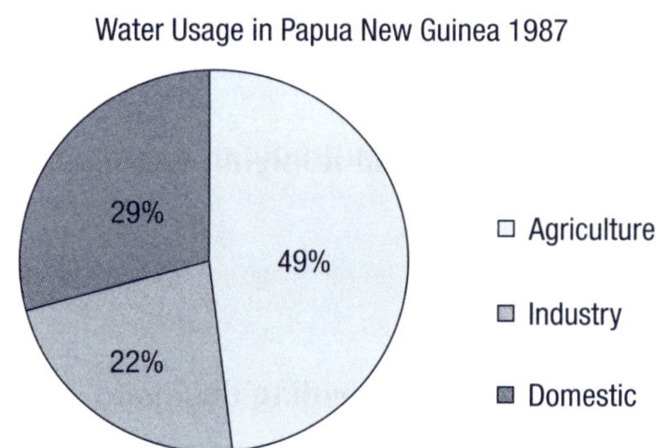

Papua New Guinea's waterways support an abundance of fish. This line graph shows the quantity (in metric tonnes) of fish caught in Papua New Guinea over a 30-year period.

4 Write a few sentences to explain what you understand from the information contained in the line graph.

5 List some of the places you have been fishing and the type of fish you caught.

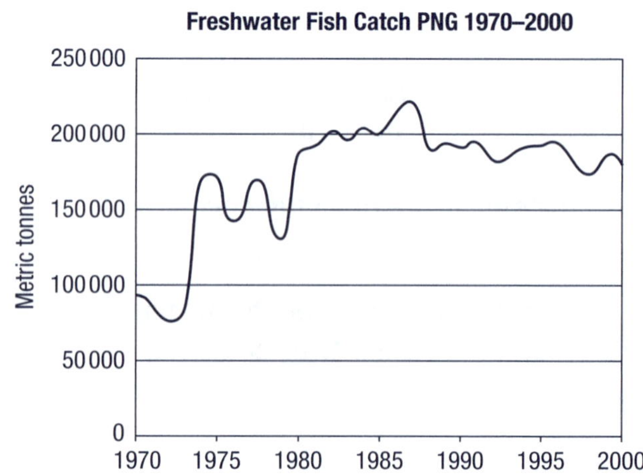

Lesson 2 Understanding, comparing and ordering decimals

The decimal number system uses place value to show the value of each number. The whole number 436 represents 400 + 30 + 6. It can be written into a place value table as follows:

Hundreds	Tens	Ones
4	3	6

The abacus below shows another way that the number 436 can be represented.

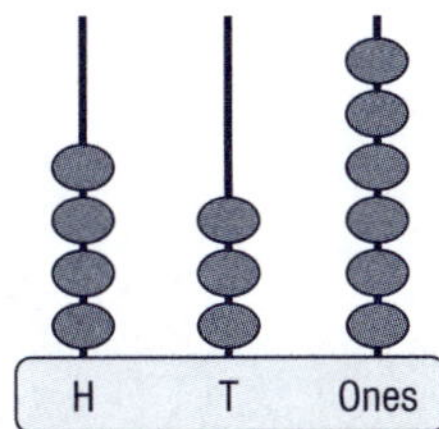

The same principles apply when writing numbers that include fractions. The whole numbers are separated from the fraction by a decimal point.

The number 5.237 represents $5 + \frac{2}{10} + \frac{3}{100} + \frac{7}{1000}$. It can be written in the place value table as:

Ones	•	Tenths	Hundredths	Thousandths
5	•	2	3	7

The abacus below shows how the decimal number 5.237 can be represented on the abacus by including a decimal point.

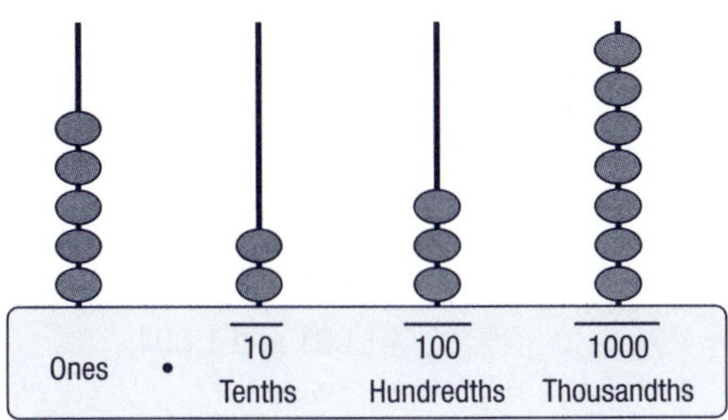

1 Draw an abacus to represent each of these decimal numbers.

a 1.24 b 2.163 c 4.201
d 0.341 e 0.019

2 What is the value of the digit 6 in the following numbers?

a 6.04 b 3.26 c 0.608
d 47.276 e 5.60

3 Write a decimal number to represent these numbers.

a $7 + \frac{6}{10} + \frac{3}{100}$ b $5 + \frac{7}{100}$
c $\frac{2}{10} + \frac{9}{100}$ d $\frac{9}{10} + \frac{6}{100} + \frac{7}{1000}$
e $2 + \frac{9}{10} + \frac{2}{1000}$ f $\frac{8}{10} + \frac{9}{1000}$
g $10 + \frac{9}{100} + \frac{9}{1000}$ h $\frac{9}{10} + \frac{8}{100} + \frac{9}{1000}$
i $17 + \frac{24}{1000}$

4 Write these decimal numbers as fractions.

a 2.7 b 4.36 c 0.427
d 2.09 e 5.003 f 2.049
g 0.049 h 8.209

In grade 6 you learned that the symbol < means 'less than' and that the symbol > means 'greater than'.

5 Copy the following table into your book and use the symbols < or > to make true statements. The first one has been done to help you.

2.5	<	2.75
0.28		0.9
2.333		2.04
1.012		1.120
0.55		0.6
3.078		4.1
3.27		3.072
1.005		1.101

6 Three rivers flow at different volumes. The river with the greatest flow flows at 5600.7 cubic metres per second. The river with the least flow flows at 5348.9 cubic metres per second. List five possible rates of volume of the other river.

7 The mouth of the Sepik River is 1.6 km wide, while the mouth of the Fly River (made up of shifting islands and tributaries) is 64 km. How much wider is the mouth of the Fly River than the mouth of the Sepik?

8 The Fly River is estimated to be 1200 km long. How long might it actually be?

9 A six month rainfall diary is shown in the table below.

Jan	Feb	March	April	May	June
15.25 cm	15.5 cm	14.9 cm	11.4 cm	11 cm	13.8 cm

a Which month had the most rain?

b Which month had the least rain?

c Arrange the monthly rainfall amounts from most to least.

10 A dripping tap wastes between 3 and 3.5 L of water each hour. List five different amounts that the tap might actually waste in one hour.

11 Copy the following numbers into your book and circle the larger number in each pair.

a 2.5 and 2.75
b 3.27 and 3.3
c 5.07 and 5.7
d 10.35 and 10.354
e 1.009 and 1.9
f 14.356 and 14.363

12 Write down five numbers between the following sets of numbers.

a 24 and 28.5
b 2 and 2.2
c 10.04 and 10.1
d 2.307 and 2.31
e 13.24 and 13.25
f 1.001 and 1.004

13 Find a number exactly halfway between the following sets of numbers.

a 3.0 and 3.1
b 1.5 and 1.6
c 2.3 and 2.4
d 34.07 and 34.08
e 21.03 and 22.03
f 0.4 and 0.9
g 0.7 and 0.85
h 197.6 and 198.5

Lesson 3 Rounding decimals

It is often easier to compare numbers by first rounding them to the same level of accuracy.

For example: For five months, Modako kept a record of the amount of fish he caught from the river in his village: 3.57 kg, 4.76 kg, 3.28 kg, 3.08 kg, 4.7 kg. It is easier to compare these figures by rounding them to the nearest whole number.

The number 3.57 is marked on the number line below.

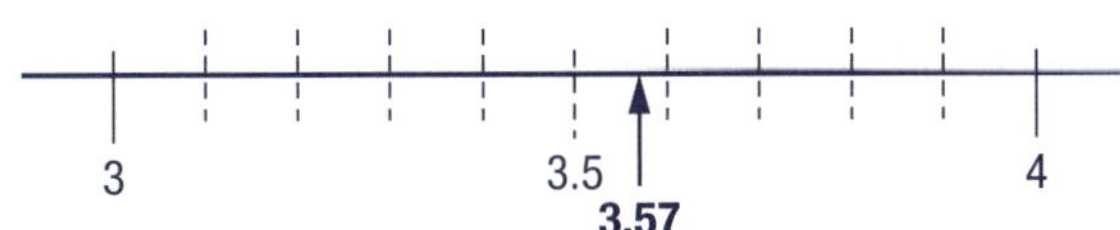

The number 3.57 is closer to 4 than to 3, so it is rounded up to 4.

1 Draw a number line to help you decide if 4.76, 3.28, 3.08 and 4.7 should be rounded up or down to the nearest whole number.

All five numbers written by Modako can also be rounded to one decimal place. The number line below shows that 3.57 is closer to 3.6 than to 3.5.

The number 3.57 rounded to one decimal place is therefore 3.6.

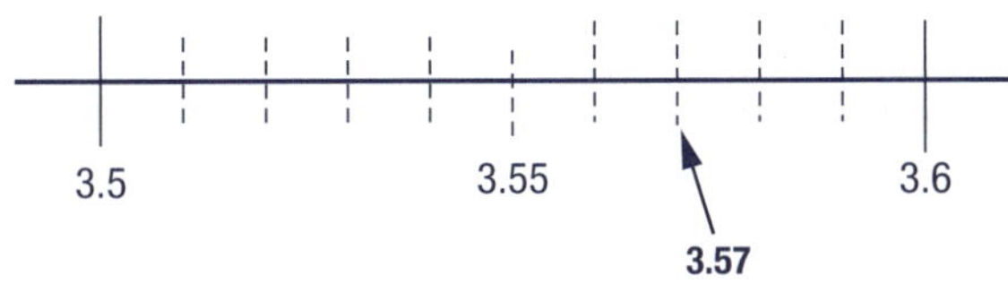

2 Draw a number line to help you decide what each of the other numbers becomes when rounded to one decimal place.

Numbers 5 and above round up, while numbers 1 to 4 round down.

For example:

5.10, 5.11, 5.12, 5.13 and 5.14 all round down to 5.1

5.15, 5.16, 5.17, 5.18 and 5.19 all round up to 5.2

3 Copy the following table into your book and complete.

Rounded to three decimal places	Rounded to nearest whole number	Rounded to one decimal place	Rounded to two decimal places
2.547			
4.866			
6.871			
7.903			

Challenge

Papua New Guinea has about 11 000 km of rivers. How long might the rivers actually be if this figure has been rounded to the nearest whole number?

4 Round the following numbers to whole numbers.

a 0.39 b 132.9 c 20.49 d 17.071

e 3.147 f 100.003 g 23.357 h 20.076

5 Round the following numbers to one decimal place.

a 3.75 b 2.35 c 27.082 d 10.02 e 4.937 f 140.30

6 Copy this table into your book and complete by using numbers that might fit the heading for each column.

Rounded to a whole number	Rounded to one decimal place	Rounded to two decimal places	Rounded to three decimal places
93	92.7		92.763
		100.36	
	4.3		
			207.983

Challenge

People often round decimal numbers when making estimations. Write down three situations when you think it would make sense to round decimal numbers to the nearest whole number.

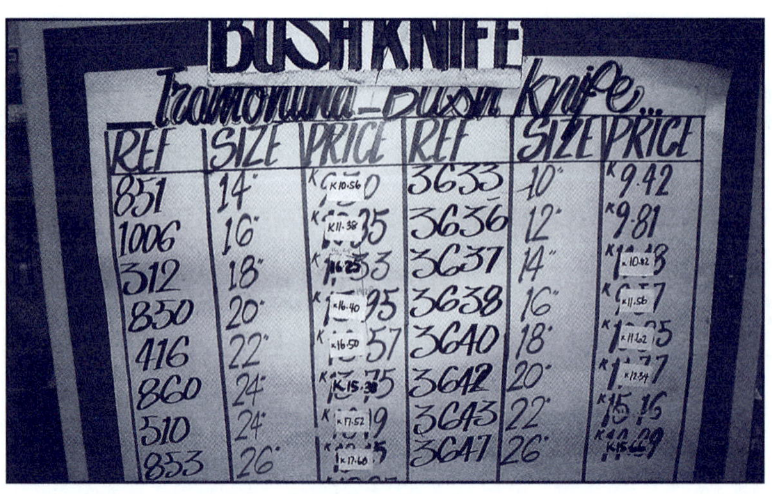

Lesson 4 Adding and subtracting decimals

Rosa is pouring water from the small containers on the right into the empty 10 L bucket. The small containers are full.

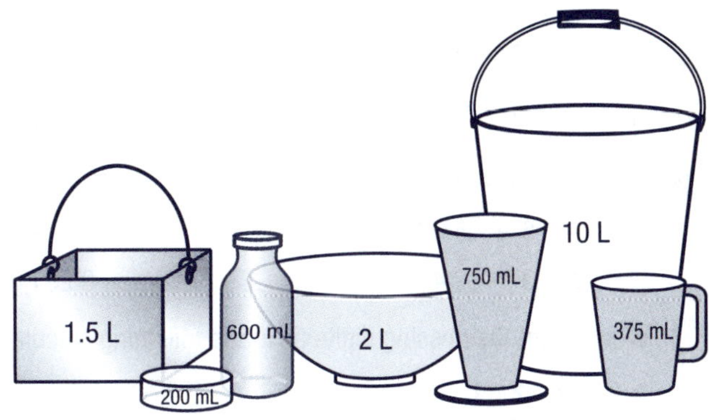

1 1.5 L is already written as a decimal number where 1.0 represents 1 L. Write the capacity of the other containers as decimals.

2 Write your answers from question 1 in order from the smallest to the largest decimal.

Help Box

When we add and subtract decimals it is important to keep the digits in columns on either side of the decimal point. It is sometimes easier to place zeros at the end of some numbers to make all the numbers in the calculation the same length.

For example:

2.73 added to 4.5 is easier to solve when written as

$$\begin{array}{r} 2.73 \\ +\ 4.50 \\ \hline \end{array}$$

34.207 minus 6.9 is easier to solve when written as

$$\begin{array}{r} 34.207 \\ -\ \ 6.900 \\ \hline \end{array}$$

3 If Rosa empties all the containers into the bucket, how much water will the bucket contain? Write your answer in mL and as a decimal number.

4 Calculate how much more water is needed to fill the bucket. Write your answer in mL and as a decimal number.

5 Sam drinks 1.5 L of water a day. His friend drinks 1200 mL.

- **a** Write 1200 mL as a decimal number.
- **b** Who drinks the most water, Sam or his friend?
- **c** What is the difference between the two quantities? Write your answer in mL and as a decimal.

6 Copy the following table into your book. Complete each column by adding or subtracting from the starting number.

0.08 less	0.2 less	Starting number	0.25 more	0.75 more
		7		
		5.3		
		9.75		
		10.06		
		30.34		

7 Make a table of your own similar to the one above and ask a friend to complete it. Be sure to make an answer sheet so that you can check your friend's answers.

8 During the wet season, the Sepik River rose 0.75 m the first week, 1.2 m further the second week and 0.9 m further the third week. How high had the river risen altogether?

9 During the dry season, Rosa's house is 4 m above the Sepik River. If the Sepik rose 2.8 m, how far from Rosa's house would it be?

10 How high above the Sepik is Rosa's house when the river level rises 3.5 m during the wet?

11 A water tank holds 200 L. If 46.8 L has been used, how much water is left in the tank?

12 Estimate answers to the following questions by rounding to the nearest whole number, then calculate the actual answer.

a 24.75 + 15.08 b 9.7 + 34.32 c 114.9 - 108.09

d 274.07 - 82.4 e 8.291 + 59.7 f 15.03 − 2.987

13 What might the missing digits be in the following calculations?

a
```
  4.2□
 −2.□□
 -----
  1.34
```

b
```
  2□.3□
 +17.□8
 ------
  □6.07
```

c
```
  □1.34
 −2□.8□
 ------
  6□.□1
```

d
```
  4□.□5
 +2□.9□
 ------
  7□.1□
```

14 Do question 13 again and find out if a different answer is possible for any of the four examples.

Lesson 5 Multiplying decimals

Help Box

Multiplying decimals by 10, 100 and 1000: When a decimal is multiplied by 10 or 100 or 1000, the answer always has the same digits in the same order. For example:

$31.42 \times 10 = 314.2$

In your head, calculate $31 \times 10 = 310$. This indicates that the answer will be about 310.

Remove the decimal point and calculate $3142 \times 10 = 31420$

Locate the decimal point in the answer: 314.2

You will notice that when multiplying by 10, the digits move one place to the left of the decimal point: $31.42 \times 10 = 314.2$

1 Copy the following table into your book. Use the method described above to calculate the answers in your head as fast as you can.

Question	Calculation
Example: 27.23 × 10	27.23 × 10 = 272.3
a 209.3 × 10	
b 1.456 × 10	
c 52.65 × 100	
d 34.07 × 100	
e 29.723 × 1000	

2 If a garden needs 3.5 L of water each week, how much water will it need for the following time periods?

a 10 weeks b 100 weeks

3 A rain gauge measures 3.8 mm of rain each day for 10 days. What was the total rainfall for the 10 days?

4 A water truck transports 1086.5 litres of water every trip. How much water could be transported if the truck made 10 trips every day from Monday to Friday?

Help Box

Multiplying decimals by multiples of 10: When multiplying a decimal by other multiples of 10, first use the same method as for multiplying by 10.

For example: 21.34×20

Calculate $21.34 \times 10 = 213.4$

Then double the answer because 20 is double 10: $213.4 \times 2 = 426.8$

Another method involves using multiplication by 100 as a first step.

For example: 15.84×50

Calculate $15.84 \times 100 = 1584$

Then halve the answer because 50 is half of 100: $1584 \div 2 = 792$

5 A village wants to install an emergency water tank that will hold at least 3.5 L of water for each of its 50 people. What is the smallest capacity tank the village should buy?

6 On average, 6.27 mm of rain fell on a town each day during April. How much rainfall might the town have had for the month?

7 Calculate the following.

a 32.41×20 b 23.04×30

c 22.31×200 d 4.12×300

e 54.027×20 f 18.302×30

g 43.025×50 h 16.231×50

8 A drink costs K1.35.

a How much for 20 drinks?

b How much for 90 drinks?

Challenge

Use what you know about decimals to solve the following questions:

How much for 15 drinks at K1.35 each?

How much for 25 drinks?

How much for 200 drinks?

Describe the method you used for each question.

Help Box

Multiplying decimals by whole numbers other than 10: When multiplying decimals by whole numbers other than 10 or multiples of 10, it is important to estimate the answer first.

For example: 38.7×16

Estimate by rounding to $40 \times 16 = 640$.

This indicates that the answer will be about 640.

Remove the decimal point and calculate:
$387 \times 16 = 6192$

Locate the decimal point in the answer: 619.2

$38.7 \times 16 = 619.2$

9 A sports team has 14 drink bottles that each holds 1.25 L. How much water is needed to fill all the drink bottles?

10 One can of cola costs K1.55. How much will it cost to buy a dozen cans?

11 Calculate the following.

a 17.6×14 b 35.7×18
c 41.9×23 d 7.13×31
e 25.06×17 f 11.803×14
g 2.095×21 h 152.39×28

Help Box

Multiplying decimals by decimals: Sometimes we need to multiply decimals by other decimals.

For example: 2.3×12.6

Step 1: Round the decimal numbers.

2.3 to 2 12.6 to 13

Step 2: Estimate the answer using the rounded numbers.

$2 \times 13 = 26$, so the final answer will be about 26.

Step 3: Multiply the numbers, ignoring the decimal points.

$$\begin{array}{r} 12.6 \\ \times\ 2.3 \\ \hline 378 \\ 2520 \\ \hline 2898 \\ \hline \end{array}$$

Step 4: Check your answer against your estimation. For the answer to be close to 26, insert the decimal point after 28 to make it 28.98.

You will notice that when multiplying decimals by decimals, that whatever the number of digits after the decimal point in the original question (in this example there were two: 3 and 6), the same number of digits will be after the decimal point in the answer (two: 9 and 8).

12 Harold has worked out that he needs 1.75 L of water for every bag of cement. He has 2.25 bags of cement left. How much water will he need?

13 Drainage piping costs K2.75 per metre. How much will it cost to buy 18.25 m?

14 Calculate the following.

a 4.9×3.5 b 18.4×2.7 c 2.31×2.5
d 7.3×1.04 e 43.2×21.25 f 19.34×12.41

Lesson 6 Dividing decimals

Dividing decimals by 10, 100 and 1000: When dividing a decimal by 10 or 100 or 1000, a similar process is followed as used for multiplication of decimals.

For example: 126.6 ÷ 10

In your head, calculate 130 ÷ 10 = 13. This indicates that the answer will be about 13.

Calculate 126.6 ÷ 10 by locating the decimal point in the answer: 12.66

You will notice that when dividing by 10, the digits move one place to the right of the decimal point: 126.6 ÷ 10 = 12.66

1 Copy the following table into your book. Use the method described above to calculate the answers in your head as fast as you can.

Question	Calculation
Example: 34.25 ÷ 10	34.25 ÷ 10 = 3.425
a 14.86 ÷ 10	
b 976.37 ÷ 100	
c 274.56 ÷ 1000	
d 34.08 ÷ 100	
e 39.04 ÷ 100	

2 If the river rose 12.75 metres in 10 days, what was the average amount it rose each day?

3 A school tank currently contains 550.73 L of water. How much water per student is this if there are 100 students in the school?

4 Fabian spent K9.75 to buy 10 fish. What was the average cost of each fish?

Challenge

Tina has collected 3.5 L of water to give to her 10 chickens. How much water can she give to each chicken? Write your answer in mL and as a decimal fraction of one litre.

Help Box

Dividing decimals by multiples of 10: When dividing a decimal by other multiples of 10 it is important to estimate the answer first. Then use the same method as for dividing by 10.

For example: 37.8 ÷ 20

Estimate 40 ÷ 20 = 2. This indicates that the answer will be about 2.

Calculate 37.8 ÷ 10 = 3.78

Then halve the answer because dividing by 20 produces an answer that is half the value of dividing by 10.

37.8 ÷ 20 = 1.89

Check your answer against your estimation.

Another method involves dividing the number by 10, 100 or 1000 first.

For example: 143.96 ÷ 300

In your head calculate 143.96 ÷ 100 = 1.4396

Then divide the answer by 3 because 100 is 3 times less than 300:

1.4396 ÷ 3 = 0.4799

Therefore 143.96 ÷ 300 = 0.4799

5 A group of friends caught 38.75 kg of fish. They decided to share the catch equally between the 20 families in the village. What amount will each family receive?

6 During April, the village installed 474.3 m of irrigation channels. What was the average length of the channels installed each day?

7 Calculate the following.

a 43.76 ÷ 20 b 97.02 ÷ 20 c 351.62 ÷ 200 d 6.24 ÷ 300

e 84.39 ÷ 30 f 18.302 ÷ 40 g 39.025 ÷ 50 h 18.405 ÷ 50

Challenge

A water container holds 35.4 L

How many 300 mL bottles can be filled from the container?

How much more water will be left in the container?

How much more water would be needed to fill one last bottle?

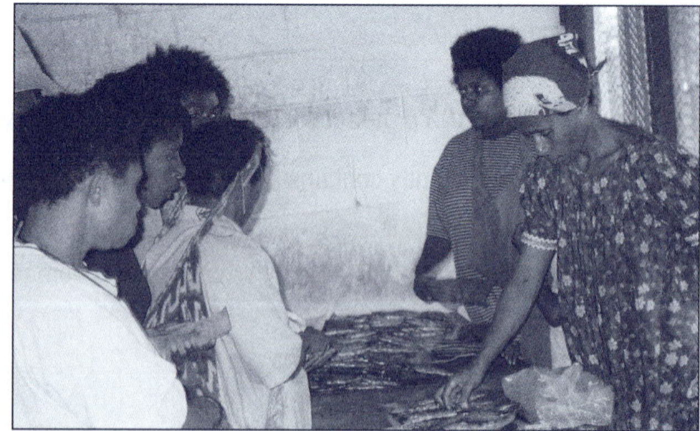

Help Box

Dividing decimals by whole numbers other than 10: Sometimes we need to divide decimals by whole numbers other than 10 or multiples of 10.

For example: 7.41 ÷ 3

Step 1: Estimate the answer first by rounding the decimals to whole numbers.

7 ÷ 3 is between 2 and 3, therefore the answer to 7.41 ÷ 3 will be between 2 and 3.

Step 2: Now calculate the division by placing the decimal point at the place where it appears during the division.

a $\begin{array}{r} 2 \\ 3\overline{)7.{}^{1}41} \end{array}$ b $\begin{array}{r} 2.\ 4 \\ 3\overline{)7.{}^{1}4\,{}^{2}1} \end{array}$ c $\begin{array}{r} 2.\ 4\ 7 \\ 3\overline{)7.{}^{1}4\,{}^{2}1} \end{array}$

so 7.41 ÷ 3 = 2.47

Step 3: Check your answer against your estimation.

The same method can be used when dividing decimals by two-digit numbers.

For example: 24.35 ÷ 18

Step 1: Estimate the answer. 24 ÷ 18 is between 1 and 2, therefore the answer to 24.35 ÷ 18 will be between 1 and 2.

Step 2: Now calculate the division by placing the decimal point at the place where it appears during the division.

a $\begin{array}{r} 01 \\ 18\overline{)24.35} \\ -18 \\ \hline 6 \end{array}$ b $\begin{array}{r} 01.3 \\ 18\overline{)24.35} \\ -18 \\ \hline 6\,3 \\ -5\,4 \\ \hline 9 \end{array}$ c $\begin{array}{r} 01.35 \\ 18\overline{)24.35} \\ -18 \\ \hline 6\,3 \\ -5\,4 \\ \hline 95 \\ -90 \\ \hline 5 \end{array}$

so 24.35 ÷ 18 = 1.35 (to two decimal places)

Step 3: Check your answer against your estimation.

8 Nicholas wanted to share 13.5 L of water equally between 8 of his plants. What amount of water will each plant receive? Write your answer as a decimal.

9 Fourteen students had together saved K34.55 to spend on an outing. How much will each student have if they receive an equal share?

10 Estimate then calculate the answers to the following (to two decimal places).

a 89.6 ÷ 3 b 59.82 ÷ 6

c 140.72 ÷ 8 d 38.04 ÷ 5

e 63.57 ÷ 12 f 24.87 ÷ 18

g 32.945 ÷ 23 h 321.45 ÷ 32

11 The village had saved K633.25 to buy a small pump. Each family had donated K37.25. How many families had donated for the pump?

Help Box

Dividing decimals by decimals: Sometimes we need to divide decimals by other decimals.

For example: 24.36 ÷ 1.2

Step 1: Round the numbers and estimate the answer.

24 ÷ 1 is 24, so the answer will be about 24.

Step 2: Remove the decimal point from both numbers.

2436 ÷ 12

Step 3: Calculate the answer to 2436 ÷ 12. $\begin{array}{r} 203 \\ 12\overline{)2436} \end{array}$

Step 4: Check your answer against your estimation. For the answer to be close to 24, insert the decimal point after 20 to make it 20.3.

12 Fabian wants to install 34.5 m of water pipe in his garden. The pipe comes in lengths of 2.3 metres. How many lengths will Fabian need to buy?

13 Estimate then calculate the answers to the following (to two decimal places).

a 9.2 ÷ 4.6 b 134.86 ÷ 0.2 c 27.2 ÷ 8.4 d 18.42 ÷ 3.6

e 40.07 ÷ 3.4 f 32.95 ÷ 3.5 g 255.4 ÷ 2.8 h 23.97 ÷ 57

Lesson 7 Measuring rainfall

Some parts of Papua New Guinea have dry months, and so do not always have water available. Tanks can be used to store rainwater caught from the roof of buildings. The stored water can be used later, especially for drinking.

The chart below shows the annual rainfall of five towns of Papua New Guinea.

	Rainfall (mm)											
Place	**Jan**	**Feb**	**Mar**	**April**	**May**	**June**	**July**	**Aug**	**Sep**	**Oct**	**Nov**	**Dec**
Goroka	230	254	266	204	113	54	49	74	121	154	171	243
Kokoda	336	333	365	328	259	187	180	224	273	321	407	362
Pomio	235	151	226	262	449	846	1261	1195	777	463	250	237
Rabaul	230	244	256	209	129	114	104	103	94	118	173	238
Wewak	139	119	154	188	227	198	189	161	207	228	196	136

1. What month has the highest rainfall in Kokoda?
2. What month has the lowest rainfall in Rabaul?
3. What month has the highest rainfall in Wewak?
4. What town has the highest monthly rainfall?
5. What town has the lowest monthly rainfall?
6. What is the difference between the highest and lowest rainfall in Pomio?
7. What is the difference between the highest monthly rainfall in Goroka and the highest monthly rainfall in Pomio?
8. Which place is most likely to need water tanks and what months would they be most needed? Why?

Challenge

Work out the mean monthly rainfall of three of the towns listed.

The rain gauges on the right show the amount of rain in millimetres that fell on a farm for one week during February.

Mon Tue Wed Thur Fri Sat Sun

9 Create a table to show the amount of rainfall each day.

10 How much more rain fell on Wednesday than Saturday?

11 What was the total rainfall for the week?

12 What was the average rainfall per day for this week?

13 If these rain gauges related to one of the towns in the previous chart, which town do you think they might represent? Explain how you decided.

Lesson 8 Area of roofs and water catchment

People choose different sized water tanks depending on the rainfall, the area of roof available, and the amount of water that is needed. It is also important to think about the best type of material for the tank, the amount of water a family or community needs, security, and how difficult it is to transport the tank.

The area of the roof will influence the amount of rainwater that can be caught.

Help Box

To find the area of this roof:

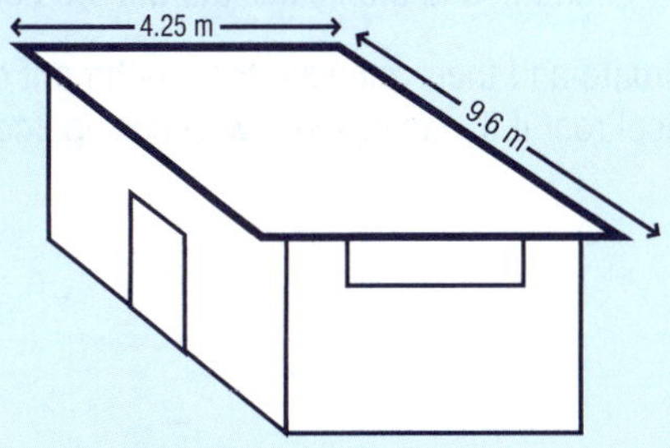

Area of a rectangle is Length × Width

So the area = 4.25×9.6

Rounding the decimals to 4×10, the area will be about 40 square metres.

Removing the decimal points, the answer to 425×96 is 40 800.

Checking against the estimate, put the decimal point back in after 40, so the area of the roof is 40.8 m^2.

1 First estimate and then calculate the area of a roof that is:

a 1.35 m long and 2.4 m wide

b 10.4 m long and 3.45 m wide

c 8.7 m long and 6.1 m wide

d 3.5 m long and 10.3 m wide

e 15.5 m long and 19.6 m wide

2 Estimate then find the area of this roof:

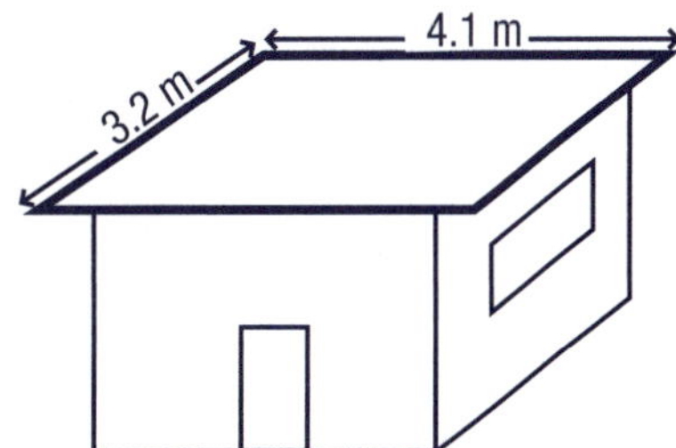

3 Copy this table into your book and complete.

Length of roof	Width of roof	Area of roof
4.5 m	2.7 m	
3.5 m	2.2 m	
2.45 m	1.3 m	
4.75 m	4.5 m	
13.25 m	5.75 m	

Challenge

A roof has an area of 12.6 square metres. What might be its dimensions?

Write another two possibilities.

4 What is the total area of this roof?

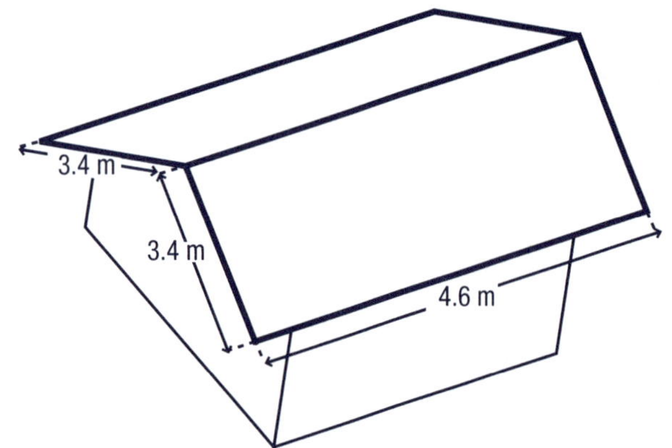

Challenge

Sometimes roofs are a composite of different simple shapes. To find the area of a composite shape you add the areas of the simple shapes together. For example, this roof is made up of a number of different sized rectangles. What is the area of this roof?

5 a A small shed measures 2.6 m by 1.7 m. The roof of the shed extends 0.5 m beyond the walls on all sides. Sketch the shed and show the measurements of the roof.

b Estimate and then calculate the area of the roof.

6 Estimate and then calculate the catchment of your school roof if it was covered with corrugated iron.

Lesson 9 Solving problems

Help Box

The formula Area = Length × Width can be used to help us work out the length or width when the area is already known.

For example:

A roof has an area of 24 m^2 and is 6 m in length.

To calculate the width of the roof, divide the area by the length: 24 ÷ 6 = 4.

The width of the roof is 4 m.

The diagram on the right shows how the area and length can be used to work out the width.

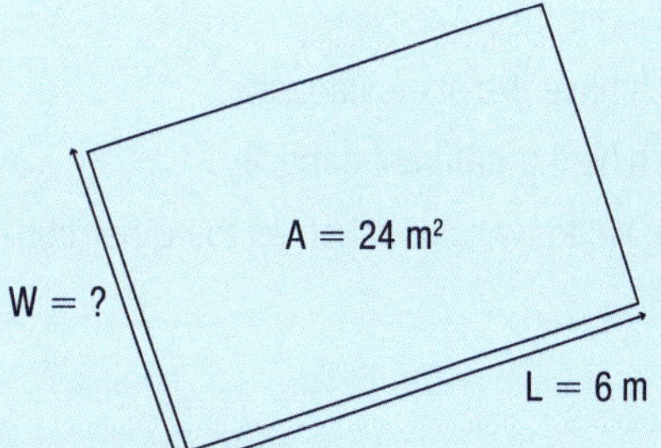

Area = Width × Length
therefore
Width = Area ÷ Length

1 How could you calculate the length of a roof if you knew the area and the width?

2 Copy the following table into your book and fill in the missing measurements.

Length	Width	Area
5 m		45 m^2
	12 m	36 m^2
3.5 m		14 m^2
	5.6 m	11.2 m^2
12.4 m	5 m	
6.2 m	8.2 m	

3 Find the width of the following roofs.

- **a** Area 40 m^2 Length 5 m
- **b** Area 36.72 m^2 Length 6 m
- **c** Area 18.6 m^2 Length 3 m
- **d** Area 50.5 m^2 Length 5 m
- **e** Area 49.14 m^2 Length 7 m

4 Find the length of the following roofs.

- **a** Area 36 m^2 Width 1.2 m
- **b** Area 7.5 m^2 Width 1.5 m
- **c** Area 18.9 m^2 Width 0.9 m
- **d** Area 40.5 m^2 Width 5.1 m
- **e** Area 50.25 m^2 Width 2.5 m

Challenge

The area of a roof is 32 m^2.

What might be the length and width of the roof?

Suggest three other possibilities.

How many other possibilities might there be?

5 Peter's roof has an area of almost 500 square metres. What might be the dimensions of the roof? Give at least three different answers.

6 In order to fill a water tank, a farmer has been told she needs to catch water from a roof with an area of between 10 and 11 square metres. Draw and label the dimensions of three different roofs that would meet this requirement.

7 A village has three water tanks that have a combined capacity of 6754 L. How much might each tank hold when full if:

- a they each have the same capacity?
- b they each hold a different capacity?
- c one tank holds twice as much as the other two combined?

Lesson 10 Calculating the amount of water to be caught

Help Box

For each 1 mm of rainfall, 1 square metre of roof will catch 1 L of water.

1 Copy this table into your book and complete.

mm of rainfall	m^2 of roof catchment	litres caught
1 mm	1 m^2	1 L
2 mm	1 m^2	
3 mm	1 m^2	
1 mm	2 m^2	2 L
2 mm	2 m^2	
3 mm	3 m^2	
10 mm	1 m^2	
20 mm		80 L
30 mm		150 L
30 mm	12 m^2	
	24 m^2	

2 Nancy needs about 20 L of water for her corn garden each week. She decides to put a small roof over her compost heap which will also collect water to feed a tank near the corn patch. What size roof do you think that Nancy should build if she lives in an area that averages about 10 mm of rain each week throughout the year?

This table shows the average rainfall for January and July in two towns.

Town	Average January rainfall	Average July rainfall
Port Moresby	178 mm	26 mm
Lae	274 mm	507 mm

3 How much water could be caught from a 12 m^2 roof:

a in January in Port Moresby?
b in July in Port Moresby?
c in January in Lae?
d in July in Lae?

4 Port Moresby has a definite dry season with an annual average rainfall of about 1000 mm. What is the maximum amount of water that could be collected from a 12 m^2 roof for a year, if it was located in Port Moresby?

5 How does this amount compare to the annual amount that could be caught if the roof was located in Lae which has an average annual rainfall of about 4600 mm?

Below is a diagram of a chicken shed. Use the information on the diagram to answer the questions.

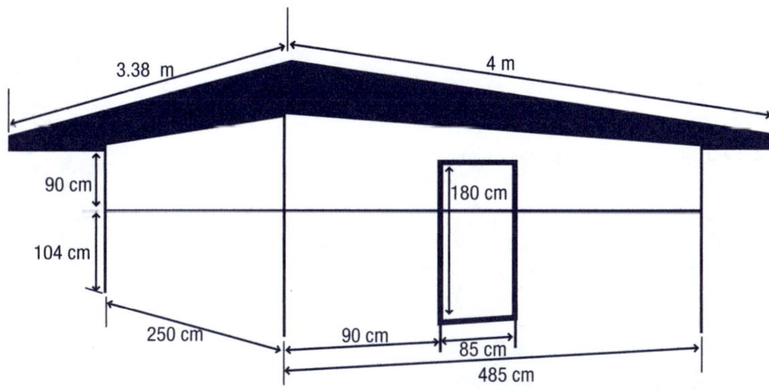

6 What area is the roof of this chicken shed?

7 How many litres of water could be collected from it if 10 mm of rain fell?

8 How much rain would have to fall on the roof to fill a 100 L water tank?

Challenge

If we put a tank on the school, how much would be collected in an average month of rainfall?

Learning Unit 3 Below the Ground

Strand: Number and Application

Ratios and Rates	Outcome 7.1.6	Recognise and relate rates to graphs
Directed Numbers	Outcome 7.1.7	Use directed numbers in concrete problems

Strand: Space and Shape

Maps and Coordinates	Outcome 7.2.16	Use a four-quadrant number plane

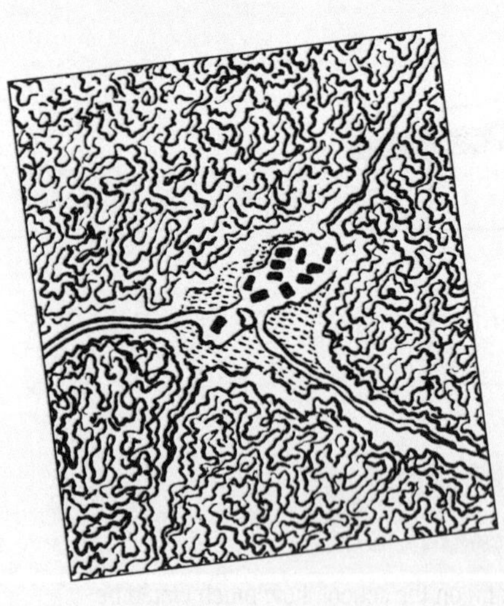

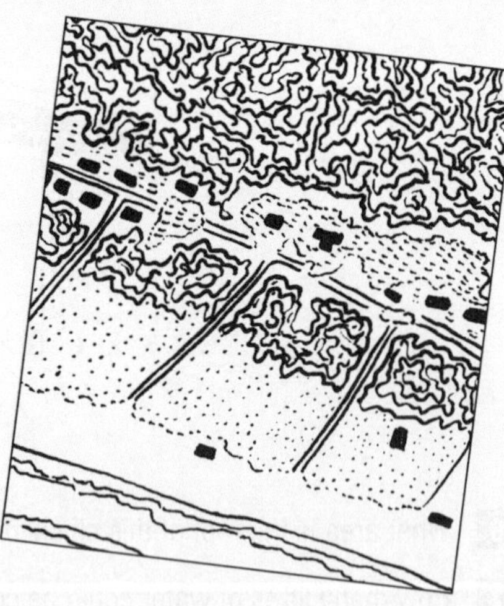

Lesson 1:	**Introduction**	Establishing the need for negative numbers in a mining context
Lesson 2:	**Directed numbers and the number line**	Describing negative numbers as opposite to positive numbers Locating integers on the number line Ordering integers
Lesson 3:	**Adding directed numbers**	Using number lines to add directed numbers
Lesson 4:	**Subtracting directed numbers**	Using number lines to subtract directed numbers
Lesson 5:	**Solving problems with directed numbers**	Solving mining related problems
Lesson 6:	**The four-quadrant number plane**	Introducing the Cartesian number plane Using ordered pairs to describe and plot points on a number plane
Lesson 7:	**Linear relationships and slope**	Completing function tables Reading and drawing linear graphs Describing gradients as positive, negative or zero
Lesson 8:	**Ratio, rise and run**	Using ratio and rate to show the relationship between two items

Lesson 1 Introduction

This unit will focus on some of the rich mineral deposits found below ground level in Papua New Guinea. You will learn about negative numbers and their place on the number line. You will also learn about rates and four-quadrant grids.

Papua New Guinea is rich in mineral deposits, and a large percentage of the money earned from exports comes from gold, copper and silver mining. Papua New Guinea also drills for and exports oil. The map shows the locations of the main deposits. Mining involves digging mine shafts and/or developing large open cut mines in order to reach the minerals that lie below the ground. Drilling for oil occurs both on and off-shore.

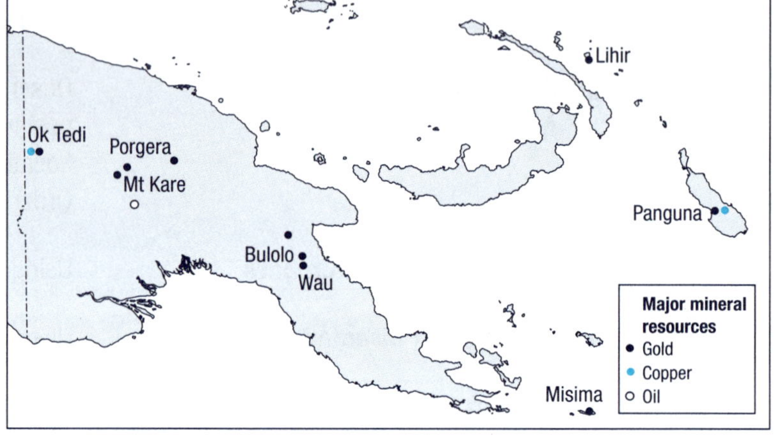

Look at the map to answer the following questions.

1 Name three places where gold is mined.

2 What is the location of the main oil deposit?

3 Which of these mines is closest to where you live?

4 What is 'open-cut' mining and how does it differ from 'shaft' mining?

Mining is one example of a real-life situation that uses numbers less than zero. Negative numbers have a value less than zero. They are identified by the negative (-) symbol in front of them.

5 List other real-life situations where negative numbers are used. Explain why each of these situations needs to have negative numbers.

On the right is a sketch of an island profile. Answer the questions by reading the number line beside the sketch.

6 How far above sea level are the following?

- a village
- b top of the mountain
- c bridge

7 How far below sea level are the following?

- a shipwreck
- b main oil deposit

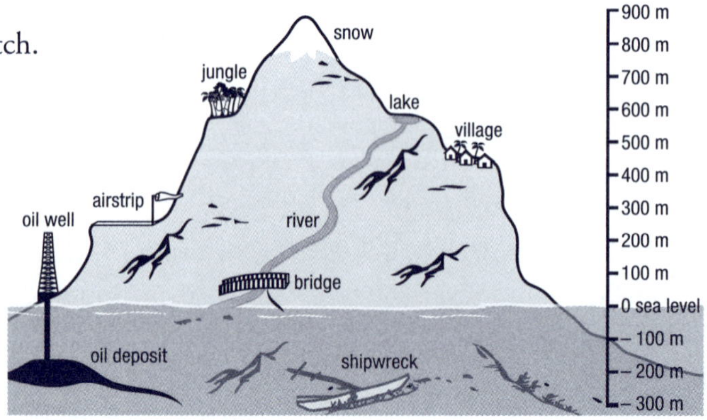

8 List three other features on the map and say how far above or below sea level they are located.

Lesson 2 Directed numbers and the number line

Large–small, up–down, north–south, profit–loss, before–after, above–below: each of these pairs of words are opposites. In mathematics, numbers can also have opposite values. Negative numbers can be described as the opposite of positive numbers. For example, a temperature of above zero is written as a positive number such as $+10°C$. The opposite temperature below zero is written as the negative $-10°C$.

Positive and negative whole numbers are called 'directed numbers' because they describe the direction of the number as well as its size. Directed numbers are often used in engineering, business and science as well as in mathematics.

1 Write the opposite of the following.

- **a** 5 km over the speed limit
- **b** K45 profit
- **c** losing 3 kg in weight
- **d** running 2 km more than yesterday
- **e** 30 m above ground
- **f** 10 seconds after blast-off
- **g** winning by 20 m
- **h** going up 9 steps
- **i** a loss of K40
- **j** 14°C above freezing
- **k** subtract 12
- **l** add 9

2 The mathematical symbols $<$ and $>$ are the opposite of each other. List other pairs of mathematical symbols that are opposites.

3 Write a directed number to show the following situations.

- **a** going up three floors
- **b** 7°C below zero
- **c** a profit of K25
- **d** gaining 5 kg in weight
- **e** 2 m below sea level
- **f** winning 3 points
- **g** 5 seconds before blast-off
- **h** losing K10
- **i** 30°C above zero
- **j** going down 5 floors
- **k** gambling loss of K20
- **l** 3 days late

Directed numbers are also known as 'integers'. A number line can be helpful when working with integers.

On a horizontal line the whole numbers to the right of zero are positive and the whole numbers to the left are negative. The further to the right that a number appears on a horizontal number line, the higher its value.

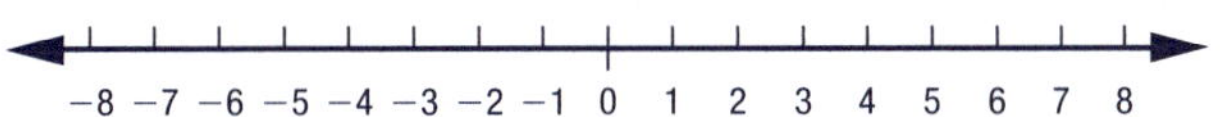

On a vertical number line the whole numbers above the zero are positive and the whole numbers below the zero are negative. Often the + sign is not used when writing positive numbers. It is accepted for example, that 5 has the same value as +5. The higher a number appears on a vertical number line, the higher its value.

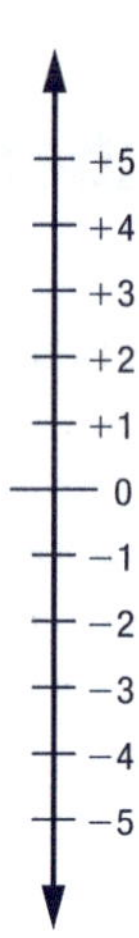

4 a Draw a vertical number line with equal spacing between marks to show the integers from −12 up to +12.

b Circle each of the following numbers on your number line: −2, +8, −7, 0, 9, −6, +3, 11, −9

c Now write the above numbers in order from lowest to highest value.

5 Copy this table into your book and write < or > between the numbers to make the statement true. The first one has been done as an example. Use either the horizontal or the vertical number line to help you decide your answers.

+7	>	+3
+4		−2
−4		−5
−1		+1
−4		+3
−5		−4
+2		−3
−2		0

6 The arrows at the ends of the number line mean that the number line continues in both directions. Imagine a longer number line and after copying these pairs into your book, write < or > between them to make each statement true.

a −45, +18
b +75, +45
c −28, −35
d −62, 0
e +27, −39
f −98, −25
g −14, +53
h +10, −11

7 Arrange the following sets of numbers in order from lowest to highest value.

a −3, +4, 7, −8, +5
b −5, −2, +4, −3, +1
c −4, +12, +7, −5, 10
d −1, −7, 9, +2, +6
e −43, −34, 43, +34, +5
f +31, −15, −28, 0, 14
g +67, +53, −45, 27, −51
h +23, −34, +10, 0, −8
i −97, 78, −45, +28, 15

Lesson 3 Adding directed numbers

Using a number line or a sketch is a practical method for helping to solve problems involving negative numbers.

Draw a number line or a sketch to help you solve the following problems.

1 A tree is growing on the top of a cliff. Esther starts at the bottom of the cliff (−7 m) and climbs up the cliff and then 6 m up into the tree. How high is she from the bottom of the cliff?

2 Write a similar problem for a friend to solve. Make sure you write the answer as well.

Help Box

Use a vertical number line to solve 4 + −6.

Step 1: The starting point on the number line is the first number in the addition.

Step 2: Treat the addition symbol between the numbers as meaning 'and'.

Step 3: The symbol + or − on the second number tells the direction to travel (up or down).

Step 4: The second number tells how many places to move.

Step 5: The finishing place is the answer.

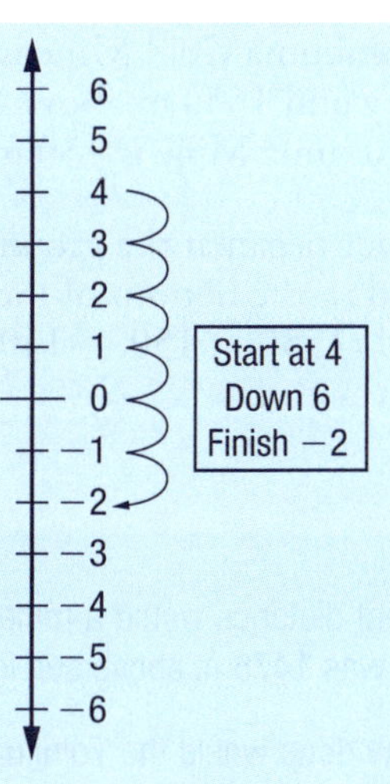

3 Draw four vertical number lines in your book from −10 to 10 and use them to solve these addition questions.

a 6 + −9 (Start at 6 and go down 9.)

b −3 + +5 (Start at −3 and go up 5.)

c 6 + 4 (Start at 6 and go up 4.)

d −5 + −4 (Start at −5 and go down 4.)

4 Copy these questions into your book and write a set of instructions (like those in question 3, above) that describe the steps needed to solve them.

a −7 + +3 b 5 + −3 c −5 + −4 d 2 + 5

5 Use a number line to solve the following.

a −5 + +8 b +7 + −4 c −2 + −3 d −6 + −4 e +2 + +9

f 9 + −2 g −4 + +5 h +5 + −5 i −12 + +6 j +4 + −8

Help Box

Horizontal number lines can be used to solve directed number questions in a similar way to vertical number lines.

On a horizontal number line the symbol on the second number tells whether to move right (+) or left (−).

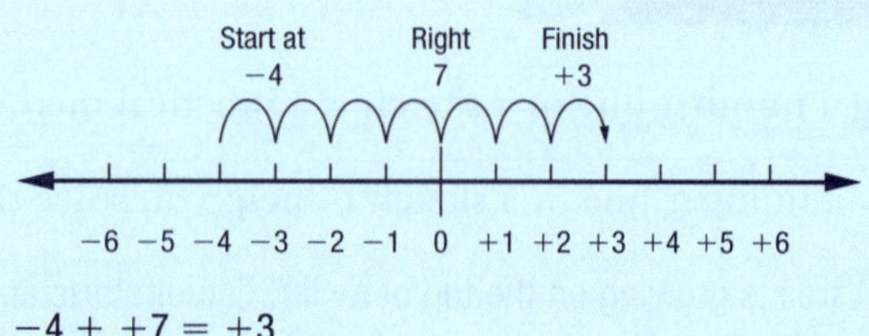

$-4 + +7 = +3$

6 Show how to solve each of the following on a horizontal number line. Write the steps used for each example.

a $-4 + +6$ b $+5 + -3$ c $-8 + -2$ d $+4 + +6$

The Tolukuma Gold Mine is located 100 km north of Port Moresby and 1550 m above sea level. The average mining depth at the Tolukuma Mine is 150 m below ground level.

If a miner began at the ground level of the Tolukuma Mine and travelled to the bottom of the mine shaft he would be 1400 m above sea level ($1550 - 150 = 1400$).

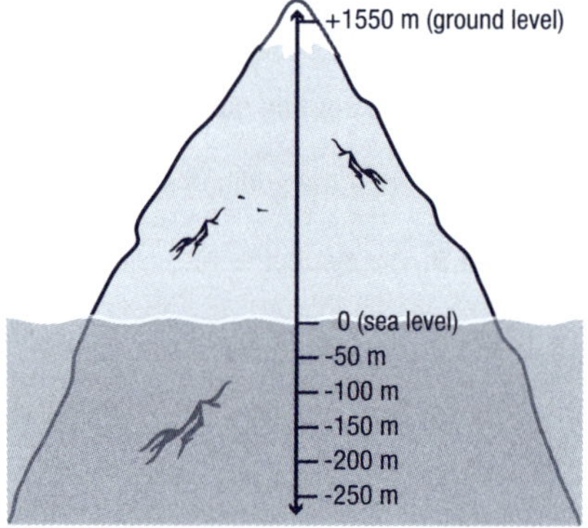

7 What distance would a miner be below the ground level at Tolukuma if he was 1476 m above sea level?

8 How deep would the Tolukuma mine have to be before the bottom of its mineshaft was 1300 m above sea level?

9 If the Tolukuma mine was made deeper, how far below the ground would a miner be in this longer mineshaft if he was 1150 m above sea level?

Most large mine shafts have a structure at ground level like the one in the picture. These structures are often a few levels high. They help with ventilation and house the lift and other equipment used to winch heavy loads up and down the mineshaft.

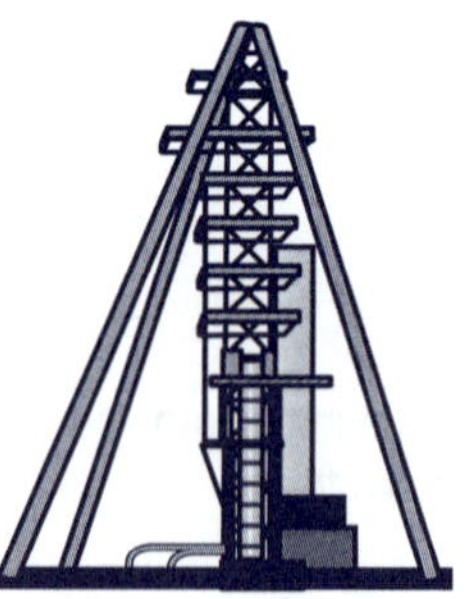

10 Draw a number line to help you answer the following questions. Write the equation you use to answer each question (eg. $-2 + -4 = -6$). What level would the lift end at if it begins at the following points?

a the 5th level below ground and moves up 8 levels

b the 8th level above ground and moves down 9 levels

c the 3rd level below ground, travels up 5 levels and then down 7 levels

d the 7th level above ground, travels 3 levels up and then 5 levels down

e the 1st level below ground, travels 5 levels up, 9 levels down and 3 levels up

Lesson 4 Subtracting directed numbers

When subtracting directed numbers it is important to remember that subtraction is the opposite of addition.

The Porgera open cut gold mine in the Enga Province, 600 km north-west of Port Moresby is considered one of the world's great gold mines.

If a truck at the Porgera mine began at level 4 below ground (−4) and drove up 2 levels, the number line would look like this:

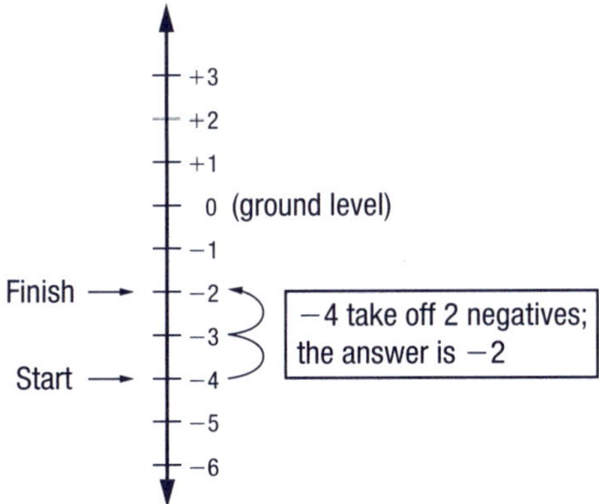

1 Draw number lines to help you decide what level the truck would finish on in the following stories. Like the example above, write a description such as '−4 take off 2 negatives means an answer of −2' beside each number line that you draw.

- **a** The truck began at −7 and drove up 3 levels.
- **b** The truck began at −4 and drove up 5 levels.
- **c** The truck began at −1 and drove up 4 levels.
- **d** The truck began at −5 and drove up 3 levels.
- **e** The truck began at ground level and drove up 4 levels.

2 Draw number lines to help you decide the answer to the following questions.

- **a** Start at −2 and take off 3 negatives.
- **b** Start at −1 and take off 4 negatives.
- **c** Start at −5 and take off 7 negatives.
- **d** Start at 0 and take off 3 negatives.

3 Write four stories of your own that involve subtracting negative numbers. Draw the number line and show how you worked out the answers to your questions.

4 Describe the direction of your movement when you 'subtract negatives'.

Mining can be extremely hot work with temperatures in deep shaft mines going as high as 60°C in some parts of the world. On average, temperature rises 1°C for every 33 metres below ground level.

So if the temperature at ground level was 25°C, the temperature in a mineshaft 132 m below ground would be 29°C.

5 Draw a number line to help you decide the temperature at the bottom of a 132 m mineshaft if the temperature at ground level was:

- **a** 15°C
- **b** 32°C
- **c** 8°C
- **d** −2°C
- **e** −7°C

6 If the surface temperature was 11°C, calculate the temperature at the bottom of the following mine shafts. Show how you calculated your answer on a number line.

- **a** 33 m
- **b** 198 m
- **c** 330 m
- **d** 462 m
- **e** 495 m

7 In April 2006, at the Beaconsfield Gold Mine in Tasmania, Australia, two miners were rescued after being trapped 925 m below ground for two weeks. Use a calculator to work out what the temperatures might have been where the two men were trapped, if the temperature at ground level was −3°C overnight and 20°C during the day.

Challenge

Use today's temperatures to calculate the maximum and minimum temperatures at the bottom of a 925 m mineshaft if there was one located in your village.

Lesson 5 Solving problems with directed numbers

At Tolukuma Mine, much recent exploration has been done using diamond core drilling. This involves a pipe encrusted in industrial diamonds being used to drill through rock to collect a 'core' of rock in the centre of the pipe. The core provides very detailed information about rock types and the mineral content.

1 A drill operator began drilling at ground level and drilled four different holes. He recorded his work on the sheet below.

Month: June 2006	Depth	Recovered	Follow up required
Site A	−158 m	clay/granite	no
Site B	−270 m	ore/copper	yes
Site C	−189 m	gold trace	yes (priority)
Site D	−25 m	limestone	no

a Which site had the deepest hole drilled?

b Which site had the shallowest hole drilled?

c What was the difference in depth between the deepest and the shallowest holes drilled?

d What was the total depth drilled in all four sites?

e The drill needs to be serviced after 1000 m. How much more drilling can be done before a service is needed if the drill had just been serviced before drilling at Site A?

2 A lift operator at the mine began his shift at ground level (Level 0). He then travelled:

down 15 levels

up 6 levels

down 4 levels

up 8 levels

down 5 levels

a What was the deepest level he travelled to underground?

b What was the highest level he reached above ground level?

c On which level did he finish?

d In what direction and how many levels would he then need to travel to return to ground level?

3 The changes in ground temperature at a snowfield were recorded every four hours. The temperature recorded at midnight was −4°C. The next five temperatures showed that at:

4 a.m. the temperature had fallen by 6°C

8 a.m. the temperature had risen by 2°C

noon the temperature had fallen again by 4°C

4 p.m. the temperature had risen by 5°C

8 p.m. the temperature had fallen by 3°C

Use a number line to solve the following questions.

a What was the temperature at noon?

b What was the lowest temperature for the day?

c What was the highest temperature for the day?

d What was the temperature at 8 p.m.?

e How far above or below zero degrees was the temperature at 8 a.m.?

4 Write a story similar to the ones above that shows what you understand about solving problems that involve directed numbers. Draw a number line to explain your story.

5 A miner was trying to pay off a credit card bill. He kept the following record of his credit card balance.

Week 1	owed K203
Week 2	paid off K25
Week 3	added an extra debt of K35
Week 4	paid off K60
Week 5	added an extra debt of K25

a How much did the miner owe on his credit card bill at the end of Week 2?

b How much did the miner owe on his credit card bill at the end of the five weeks?

c What was the largest amount that the miner owed on his credit card debt during the five weeks?

6 Copy these puzzles into your book. Complete by repeatedly carrying out the operation described in the centre circle to the surrounding numbers. Two have been done as examples.

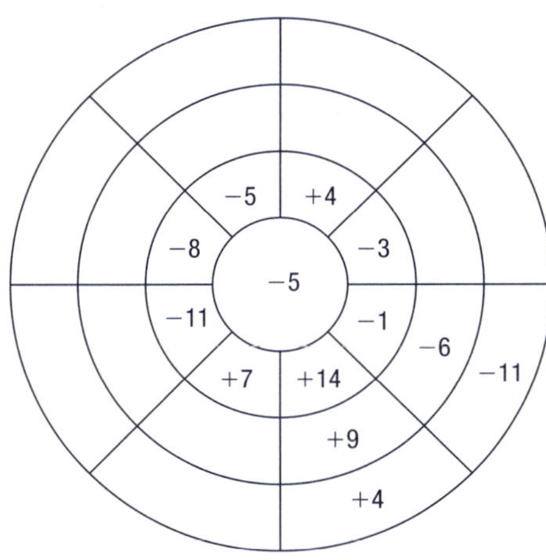

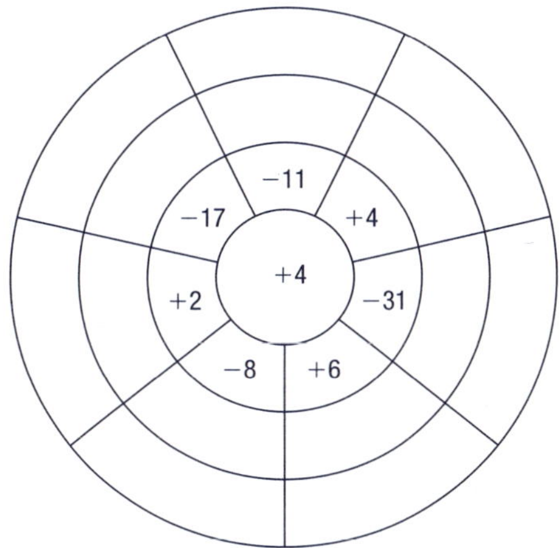

Challenge

Copy the grid into your book.

Arrange the integers −4, −3, −2, −1, 0, +1, +2 into the grid so that the sum of the three numbers along each dotted line equals −3.

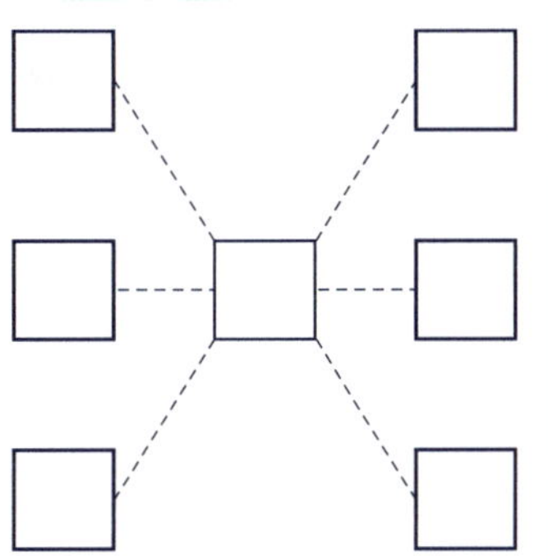

Lesson 6 The four-quadrant number plane

Any pair of numbers can be used to plot a point on a number plane. Both positive and negative numbers can be represented on the number plane.

The diagram shows how the x and y axes divide the number plane into four quadrants.

A pair of numbers (also called coordinates) is used to describe a point on the number plane.

The first number in the pair describes the distance left or right of the origin (along the x-axis).

The second number in the pair describes the distance above or below the origin (along the y-axis).

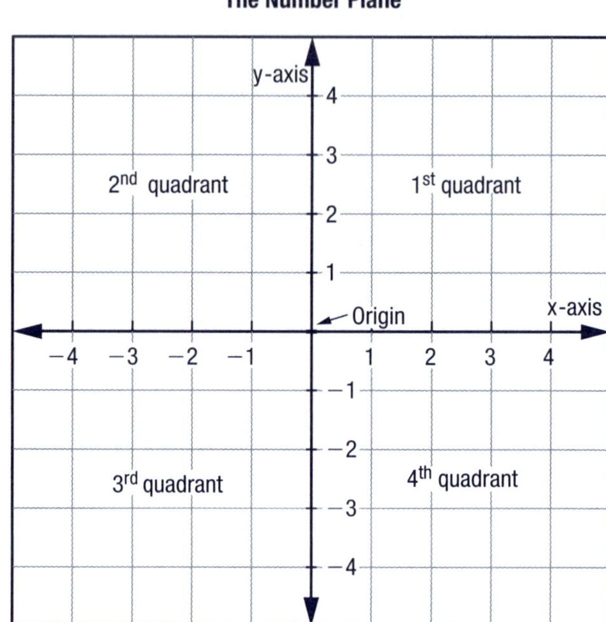

The order of the numbers is very important when plotting a point on the number plane. It is important to write the location on the x-axis first. The number plane on the right shows the ordered pair (−2, 4).

1 Give ordered pairs for the following points on the number plane.

- **a** Point A
- **b** Point B
- **c** Point C
- **d** Point D

2 Draw a number plane and plot the following ordered pairs:

- **a** (+4, +3)
- **b** (−5, +4)
- **c** (+2, −3)
- **d** (−4, +1)
- **e** (−2, −2)
- **f** (−4, 0)

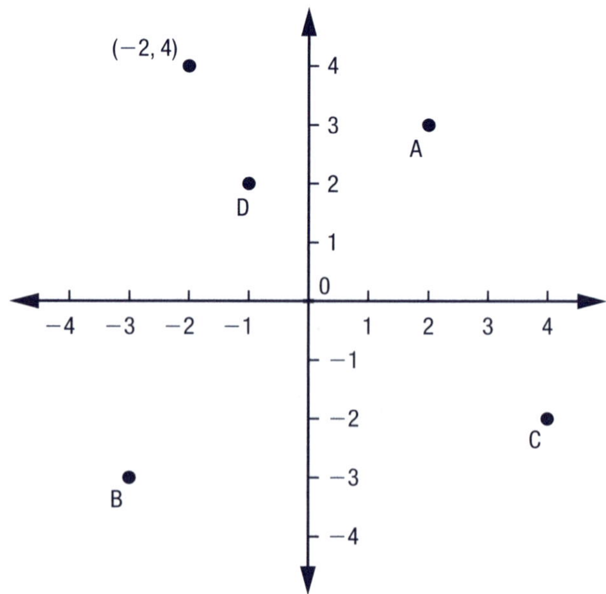

3 The air tunnels shared between two mining companies have been graphed on a number plane and emergency stations plotted.

a Write the coordinates for each of the eight emergency stations.

b How many of these emergency stations are located in the Andrew Mine (including the one on the boundary)?

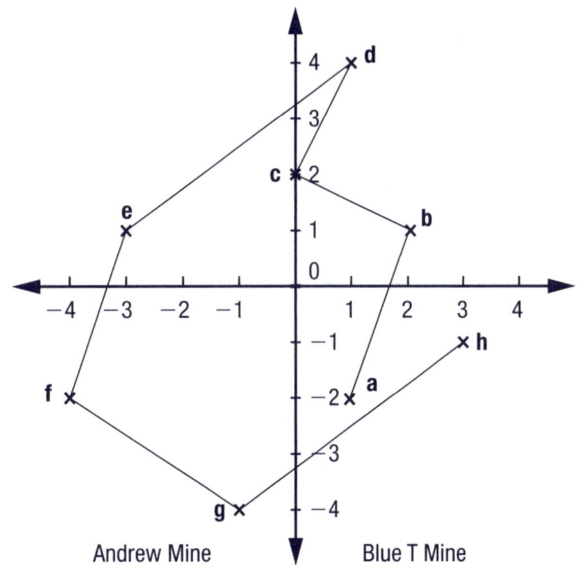

4 Rule up a number plane with a scale of −4 to +4 on both axes. Plot the following coordinates and join them in order to form a picture:

(−1, 3) (−1, 1) (−3, 1) (−3, −1) (−1, −1) (−1, −3) (1, −3) (1, −1) (3, −1) (3, 1) (1, 1) (1, 3).

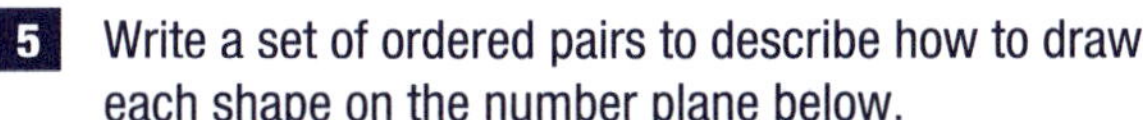

5 Write a set of ordered pairs to describe how to draw each shape on the number plane below.

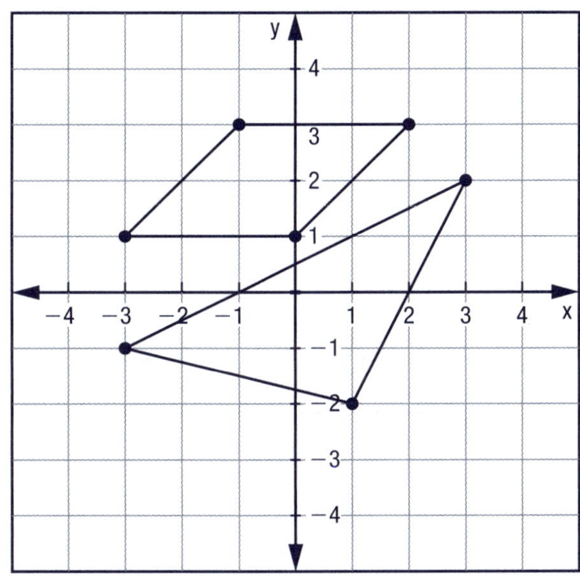

6 Name the two shapes.

7 Rule up a number plane and draw two other mathematical shapes. Write a set of coordinates to describe how to draw each shape.

Challenge

The alphabet has been written on this number plane.

Write the set of coordinates that spell your name.

Write the set of coordinates that spell your teacher's name.

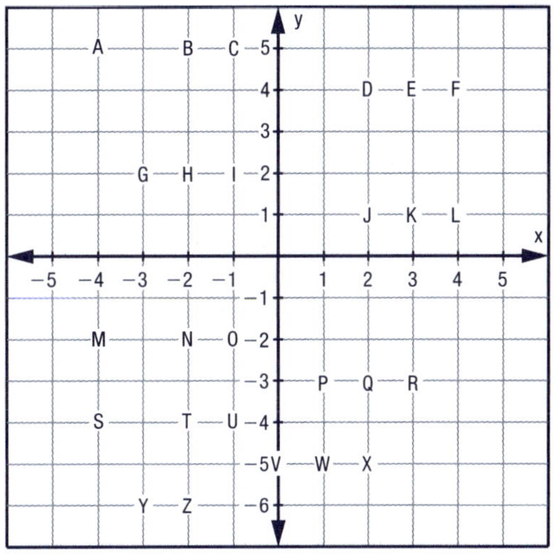

Lesson 7 Linear relationships and slope

The number plane can also be used to show relationships between different things. Earlier you learned that on average, temperature increases in underground mines at the rate of 1°C for every 33 m below the ground. The table below shows the relationship between temperature increase and distance below the ground.

Distance below the ground	33 m	66 m	99 m	132 m	165 m	198 m
Temperature increase	1°C	2°C	3°C	4°C	5°C	6°C

1 Copy and complete the following tables showing the relationship between two things.

a A mining exploration team drills down 3 metres every 5 seconds.

Depth	3 m	6 m	9 m	12 m	15 m
Seconds	5	10			

b A miner had 4 days off for every 14 days worked.

Days off	4	8			20
Days worked	14	28		56	

c The price of gold was K18 per gram.

Grams	1	10	100	500	1000
Price					

The relationship between the distance underground and the temperature increase can be graphed using 'Distance' as the x-coordinate and 'Temperature increase' as the y-coordinate.

The points on the graph can be connected to show a straight line. This means that the relationship between the two sets of numbers is a linear relationship.

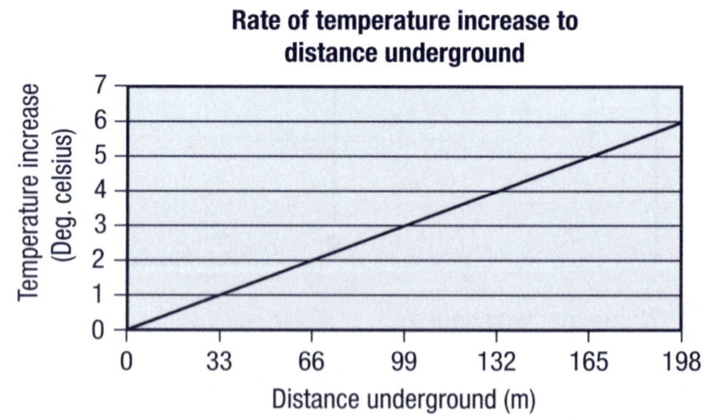

2 Look back at the first two sets of data in question 1.

a Sketch what you think each slope would look like for 1a and 1b if they were plotted on a graph.

b Explain why you think each slope would look like the sketches that you drew.

The data in the following table can be written as coordinates and then plotted on a graph. Example: (0, 0) (2, 1) (4, 2) and so on.

x-axis	0	2	4	6	8	10
y-axis	0	1	2	3	4	5

3 Draw the graph and plot the coordinates. Join the points with a straight line.

The gradient describes the steepness or slope of a line.

4 Look at these lines.

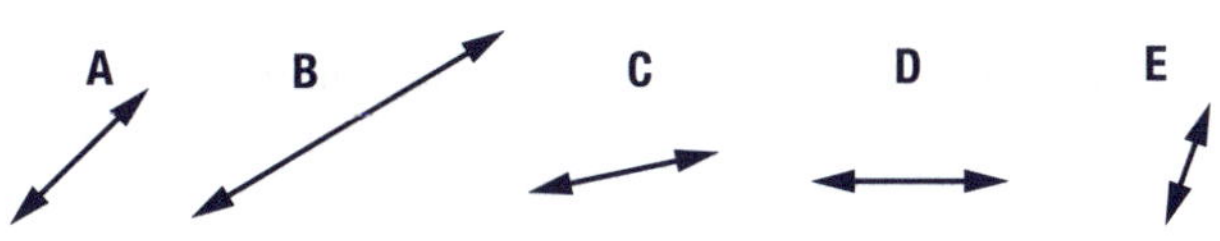

a Which line is the steepest?

b Which line has the smallest slope?

c Order the lines from steepest to horizontal.

d If the line represented a path that you had to walk, which one would require the most energy?

Help Box

The line slopes upwards from left to right. This is called a 'positive' gradient.

The line slopes downwards from left to right. This is called a 'negative' gradient.

A horizontal line has a 'zero' gradient.

5 Copy these lines into your book and describe their gradient as either positive, negative or zero.

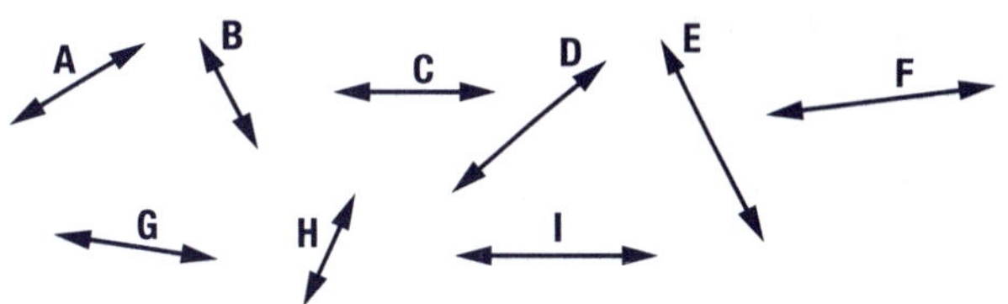

6 Below is a plan of a mine. Describe the gradient of each labelled shaft.

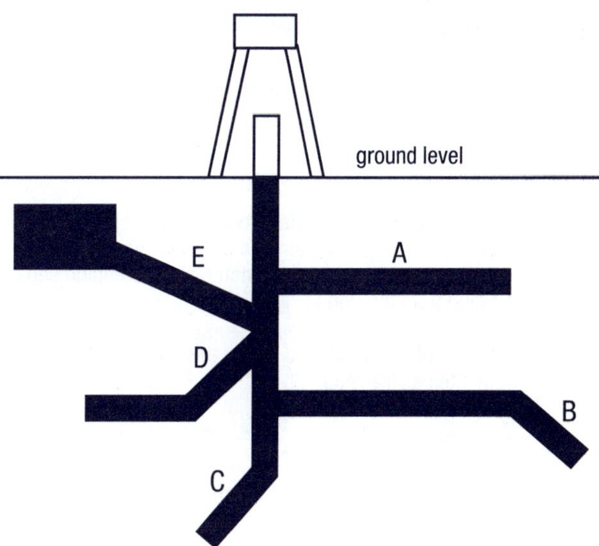

Lesson 8 Ratio, rise and run

In the previous lesson we saw how graphs show the relationship between two items. The 'Distance – Temperature increase' graph showed the rate that temperature increases in relation to distance underground.

The ratio of distance to temperature is 1:33. This can also be written as $\frac{1}{33}$.

We can also say that the rate of increase is 1:33 degrees per metre.

Help Box

We can use ratio and rate to show the relationship between two items.

For example:

Iron is extracted from certain rocks, known as iron ore. High quality iron ore can contain one part non-iron substance, known as slag, and two parts natural iron. Refining separates the slag from the iron.

The ratio of slag to natural iron in this case is 1 to 2. This can be written as the ratio 1:2.

Ratios can also be written as fractions. The fraction showing the ratio of slag to iron is $\frac{1}{2}$. The rate that slag is produced relative to the iron production is 1:2 units of slag per unit of iron.

In 2004 the Porgera gold mine sourced 12% of its gold from underground mining and 88% from open pit mining.

The ratio of gold from underground mining to gold from open pit mining was 12:88.

This can be simplified to 3:22 and written as the fraction $\frac{3}{22}$.

1 Write the ratio (in its simplest form) and the fraction for the following relationships.

- a 5 metres drilled every 30 minutes
- b 12 grams of gold for every 36 tonnes of rock
- c 6 L of water drank every 18 hours
- d 12 heartbeats per 6 seconds
- e 15 kilometres travelled in 45 minutes
- f K100 earned for every hour worked
- g 1200 kilojoules used in 4 hours

2 Write the rate and unit of measurement (e.g. degrees per metre) for each relationship in question 1.

Usually, but not always, stairs are built so that every step up is about 18 cm and the length of each step is about 27 cm.

The ratio of distance 'up' (called rise) to distance 'forward' (called run) is 18:27.

The broken line in the diagram shows how steep the steps are.

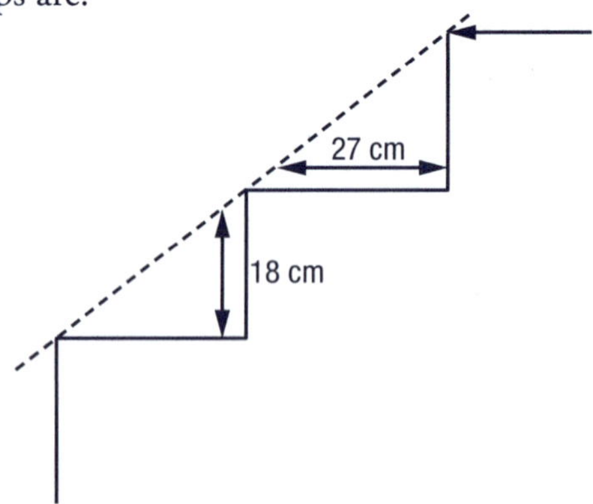

3 a Write the ratio 18:27 in its simplest form.

b Is the gradient of the broken line positive, negative or zero?

4 a Write the ratio for this stair diagram in its simplest form.

b Describe the gradient of the broken line.

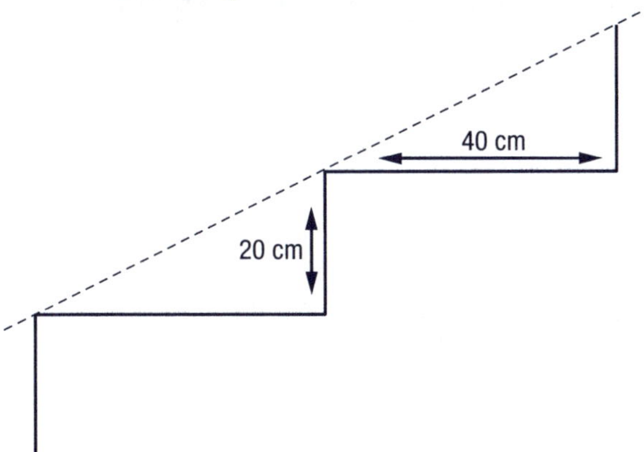

Help Box

The stairs shown on the left have been drawn to scale. The ratio has been used to help decide the length of the rise in relation to the run. It is easy to see the link between ratio, fractions and scale.

Example: The ratio 20:40 can be simplified to 1:2.

The fraction $\frac{20}{40}$ can be simplified to $\frac{1}{2}$.

Stairs measuring 10 cm (rise) and 20 cm (run) is a 1:2 ratio and can be enlarged to 30 cm (rise) and 60 cm (run).

5 a Measure any sets of steps at your school and calculate the ratio.

b Draw a graph to show how steep the steps are.

c Is the gradient of the plotted line positive, negative or zero?

Challenge

Draw two more sets of stairs to scale using different measurements.

Write the ratio of rise to run in the simplest form for each set and then describe the gradients.

6 The following graphs show the distance a miner travelled away from the main tunnel of a mine during different time intervals.

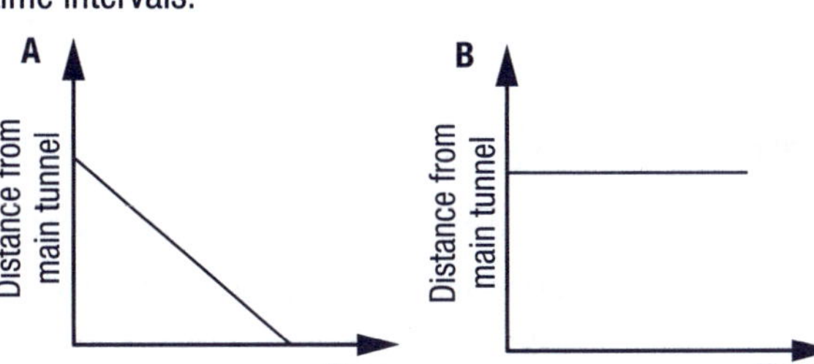

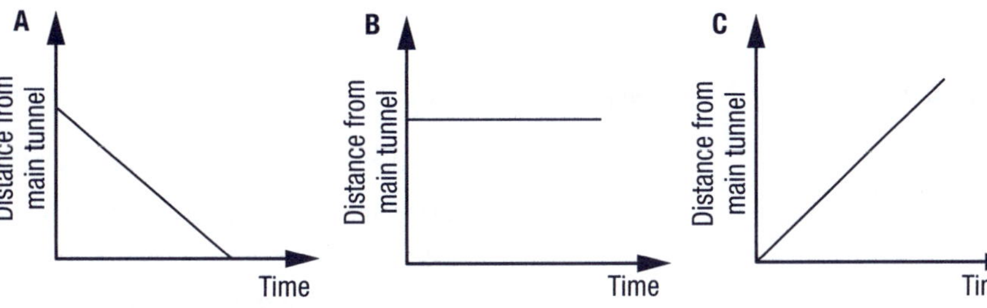

a Describe the slope in each graph.

b Which graph shows the miner moving away from the main tunnel over time?

c Describe the relationship between 'distance from the main tunnel' and 'time' for the remaining two graphs.

d Draw another graph that shows a different slope and describe the relationship between 'distance from the main tunnel' and 'time' for this graph.

7 a Predict what the slope would be like on a graph that shows the amount of petrol left in a mining truck as it drove around during the day.

b Draw a graph to show the predicted slope.

c Describe the relationship between 'time' and 'petrol remaining' as shown by the slope of your graph.

Learning Unit 4 Revision Challenges

Lesson 1 Revision

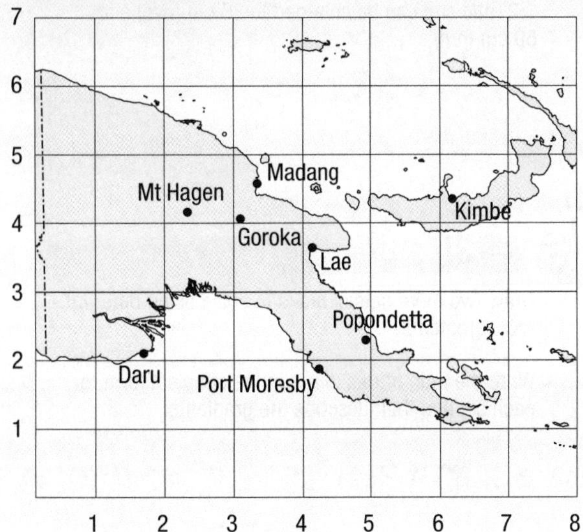

1 Write the nearest coordinates to locate the following places.

a Port Moresby b Popondetta
c Daru d Madang
e Goroka f Mt Hagen
g Kimbe h Lae

2 Follow the instructions to draw the shape.

- Mark a starting point on a piece of paper.
- Rule a 3 cm line north-east.
- Turn 90° clockwise and rule a 3 cm line.
- Quarter turn and rule a 3 cm line.
- Turn north-west and rule a 3 cm line.

3 Give the compass bearings for the location of A, B, C and D.

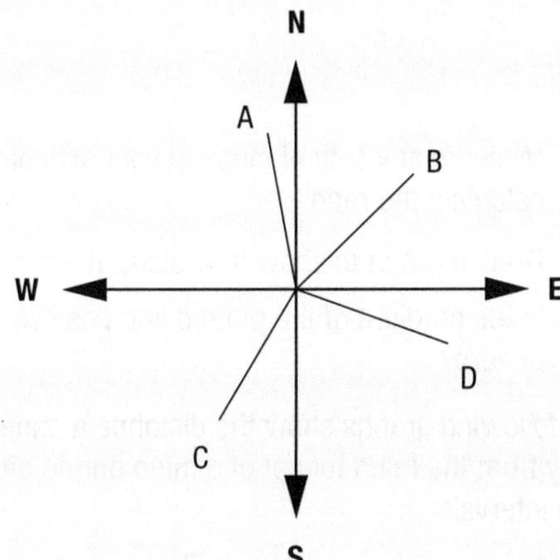

4 The scale on a map shows: 2 cm = 10 km

a How far does 5 cm on the map represent in km?

b How far would a distance of 45 km appear on the map?

c Write the scale as a ratio in its simplest form.

5 Rewrite the following scales as ratios in their simplest form.

a 1 cm = 1 km b 3 cm = 2 m

c 5 cm = 1 m d 2 cm = 10 km

6 Simplify the following ratios.

a 5 to 100 b 3:27 c $5\frac{1}{2}$:22 d 5 cm = 1 km

e $\frac{8}{32}$ f 2 cm to 10 m g $\frac{100}{1000}$ h 90 to 3

7 Enlarge the following measurements using a 2:3 ratio.

a 10 cm b 4 m c 20 mm d 4 kg e 50 mL

f 6 mm g 12 L h 8 min i 100 m j 20°C

8 Copy the table and complete the gaps.

Scale	Ratio	Fraction
1 cm = 3 m		
	2:5	
3 cm = 5 km		$\frac{3}{5}$
4 cm = 100 km		

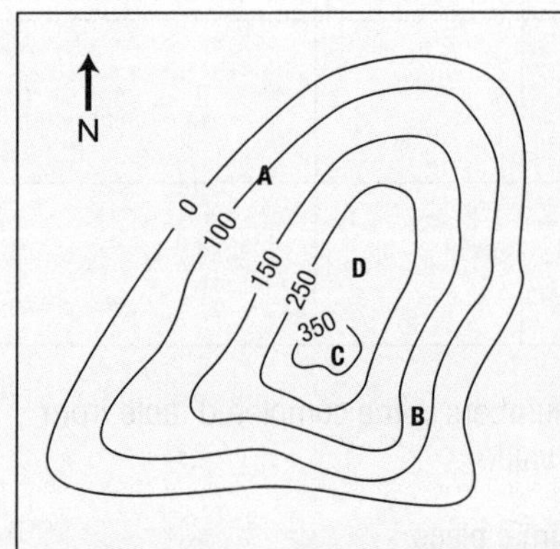

9 Use the contour map to answer the following questions:

a What is the height above sea level at location A?

b Which location is at the highest point on the island?

c Sketch a profile of what the island might look like if you were standing on a boat anchored off its west coast.

d Sketch a profile to show what the island might look like if you were standing at location A looking south.

10 The diagram below is a network map of five villages. Investigate if it is possible to travel from one village to another without visiting each village more than once.

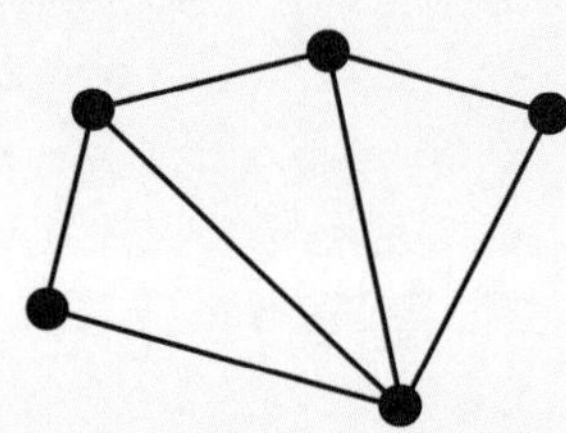

Lesson 2 Revision

1 Copy and complete the table.

Abacus	Decimal	Fraction
[abacus]	2.312	
		$4 + \frac{3}{10} + \frac{2}{100} + \frac{6}{1000}$
	3.07	
		$\frac{7}{10} + \frac{4}{100}$
[abacus]		

2 List the decimal numbers in the completed table from highest to lowest value.

3 Round to one decimal place.

a 2.375 b 4.209
c 3.27 d 4.08
e 2.05 f 1.93
g 0.06 h 0.048

4 Round to the nearest whole number.

a 2.78 b 4.079
c 5.5 d 2.07
e 1.93 f 2.85
g 4.09 h 2.509

5 Order these decimals from most to least value.

2.67, 2.76, 6.72, 6.27, 2.07, 2.6, 0.673, 0.68, 0.67, 2.06

6 Write three numbers that come between the following sets of numbers.

a 2.75 and 2.78
b 0.3 and 0.32
c 4.37 and 4.03

7 Copy and complete the table.

0.5 less	Number	0.5 more
	2.7	
	0.3	
	4.02	
	2.52	

8 Estimate first where necessary and then calculate.

a $24.3 + 3.09$ b $102.79 + 3.682$
c $10.307 - 8.92$ d $81.05 - 7.342$
e $6.93 - 4.23$ f 5.42×10
g 3.76×30 h 2.07×18
i 4.63×2.5 j 2.09×1.3
k $42.6 \div 100$ l $2.84 \div 20$
m $13.95 \div 5$ n $49.14 \div 17$
o $25.32 \div 1.2$ p $184.2 \div 0.6$

Lesson 3 Revision

1 Copy these number pairs into your book and insert $>$ or $<$ between each pair to make true statements.

a $-8, -5$ b $4, -8$ c $34, -21$ d $28, -10$ e $-1, 6$

f $2, -9$ g $-5, 0$ h $9, -16$ i $13, 14$ j $-11, -10$

2 Draw a number line and write about how it can be used to solve $-4 + 7$.

3 A large company owns a building that has 8 floors above ground level and 4 floors below ground level. An employee takes the lift to visit offices on different floors in this order:

start at ground level; go up 3 floors; go up 2 floors; go down 6 floors; go down 2 floors; go up 4 floors

a Draw a number line to work out what floor the employee finishes on.

b What was the lowest floor the employee visited?

c What was the highest floor the employee visited?

d Which floors did the employee not visit?

4 Give the ordered pair for the following points on the number plane.

a Point A b Point B

c Point C d Point D

e Point E f Point F

g Point G

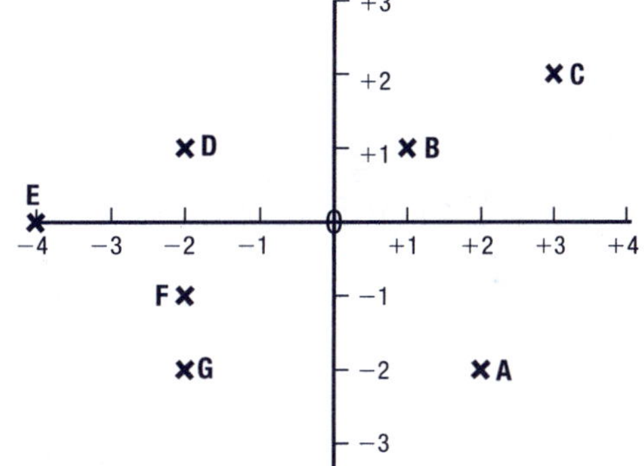

5 Draw a four-quadrant number plane and plot the following coordinates by labelling the location with the matching letter from the alphabet.

A = $(3, -2)$ B = $(-4, 5)$ C = $(1, 3)$ D= $(-2, -3)$ E = $(0, -2)$ F = $(3, 0)$

6 Draw three different straight lines that all have a positive gradient.

7 Describe the rate (in its simplest form) and unit of measurement for the following relationships.

a 50 g of sugar for every 40 kg of flour

b 40 kilometres travelled for every litre of petrol

c 2 cm per minute

d K125 for 5 hours work

e 80 kilometres travelled in 2 hours

Investigations

These investigations provide an opportunity for you to show what you understand about mapping and about working with decimals. The work you have done in the topic 'Our Country' will help you to complete these tasks.

Task 1: My island cruise

Draw a map of an imaginary island. Include the scale, direction north and some contour lines.

Use a ruler and a protractor to draw a course for a ship to sail around the island. The course needs to be made up of a series of straight lines.

List the compass direction, bearing and length of each section to be sailed.

Mark three different locations where the ship might anchor. Using the contour lines as a guide, sketch the different profiles of the island that might be seen from the ship at those three locations.

Task 2: Decimal poster

Prepare a lesson for Grade 6 where you teach the mental arithmetic rules for decimals. For example: how to multiply by 10, how to divide by 100, and so on.

You should also make a poster that explains simply, in both words and with numbers, the main rules that apply to adding, subtracting, multiplying and dividing by decimals.

Include some real life examples of when people might need to use decimal numbers in this way.

Investigations

Task 3 provides an opportunity for you to show what you understand about area and capacity. The work you have done in the topic 'Our Country' will help you to complete this task.

Task 3: Collecting rainwater for my family

Draw a bird's-eye view of your house (or your school, church or trade store if the roof of your house is bush material rather than corrugated iron). Label the diagram with your estimation of its length and width.

Sketch the roof of the house and include an estimate of its dimensions.

Calculate how much water you could catch off your roof after:

a 1 mm of rainfall

b 3 mm of rainfall

c 11 mm of rainfall

Think about how much water you use each day. Make a list of the main things you use water for during the day and write an estimate of how much water each activity takes. What is the total amount of water you use daily based on this estimate? Calculate what size tank would be suitable for your needs. (Show your working out.)

Calculate what size tank would meet the total needs of all the people living in your house.

If your house had a rainwater tank with a capacity of 600 L, what would be the best way to decide how much water each person was entitled to? If water was scarce in your area during the 'dry' season, what rules do you think your family should have about using the tank water?

Investigations

Task 4 gives you the opportunity to show what you understand about directed numbers. The work you have done in the topic 'Our Country' will help you to complete this task.

Task 4: Create a magic square using directed numbers

Complete the magic square below using these numbers: −3, −2, −1, 0, 1, 2, 3, 4, 5

The sum of each row, column and diagonal must equal +3.

Cut out nine small paper squares and write one number on each. This way you can move the numbers around in the grid until the rows, columns and diagonals make the correct total.

Create another magic square with different numbers and a different total for your friends to try. Your new magic square must use at least five numbers and must include both positive and negative numbers.

For Teachers

For Teachers

The Oxford Grade 7 Mathematics Program was developed to reflect the teaching approaches described in the Upper Primary Teachers Guide (2003) produced by the Department of Education, Papua New Guinea. The Teachers Guide provides valuable information about curriculum development, appropriate use of context, selection of suitable learning and teaching strategies and how to plan and carry out effective assessment. The Grade 7 Mathematics Program should be used in conjunction with the Teachers Guide to help teachers implement the Grade 7 Mathematics Syllabus.

The Grade 7 Program comprises four topics across two books. Book A and Book B each contain two topics and it is expected that each topic will take one term to complete.

Each topic is divided into three Learning Units and one Revision Challenges unit.

All learning outcomes for each Mathematical Strand have been covered in the Oxford Grade 7 Mathematics Program. On later pages a chart matches the learning outcomes to the various Learning Units.

A key focus of the Grade 7 Mathematics books is that they are based on particular contexts. Four broad topics of general interest have been chosen and each Learning Unit examines a specific context within these broader topics.

The mathematical concepts have been organised to build upon each other as students work through the two books. This means that in order to make learning as effective as possible, students should understand the concepts contained in Book A before beginning Book B. However, depending on your students' needs, you may decide to alter the order in which they complete the topics.

Planning your teaching

In Upper Primary 180 minutes per week has been allocated to mathematics. The National Department of Education's policy is to allow flexibility in timetabling and you may, for example, plan to teach five 36-minute lessons or four 45-minute lessons a week.

As a general guide teachers should aim to finish one topic in the Grade 7 Mathematics book each term. The individual lessons vary in length but each would be expected to take a minimum of 36 minutes to complete and in some cases may take two or three lessons. Plan to spend 2 to 3 weeks on each Learning Unit before completing some of the Revision Challenges for the topic in the final week of the term.

Using contexts

Recent reforms in education have highlighted the need to support students so they can use the mathematics they learn in realistic, practical situations. Using a context for teaching mathematics has the added advantage of helping students to see the meaning of the concepts and of making the posing of mathematical problems easier to understand.

It is important that the curriculum builds links for students between the mathematics that they do at school and its application to real-life situations.

Explain why you are using the contexts

Linking new mathematical concepts to familiar and interesting contexts enables students to integrate their learning in a way that relates more readily to their own lives. The contextual approach requires students to use knowledge and skills from a range of sources to find solutions using the methods and levels of accuracy appropriate to the real-life context.

Discuss the context with students

The context is useful for showing students how the mathematics is used in real-life. Generally, contextualising the mathematics also makes it easier to understand. However, in some cases the context can make the mathematics more difficult to learn. For example, if you want to talk about sea-shells, students from Highlands region may have difficulty understanding what you mean. It is very important that students understand the context that is being used. Teachers should adapt the language and/or the context to best meet the needs of their students.

Transfer the concept to another context

Mathematical knowledge and skills gained in one context are only useful if they can be applied in some other context. While it is helpful to initially teach a concept using a familiar context, it is important that the student finally be asked to solve a similar problem using a different context. Only then can teachers know that the concept has been learned.

Topic introduction

Each topic in the Grade 7 Mathematics Program begins with a double page spread that provides an overview of the broader context and its relevance to the student. The links between the context and the mathematical concepts for each Learning Unit also appear on this double page spread. Use the topic introduction pages to initiate discussion about the context and to find out what students already know about the real-life situations described and the mathematics involved.

The first lesson in each Learning Unit introduces the specific context of that unit within the broader topic. The introduction brings the context into the realm of the individual student and explores their existing knowledge of the context at a personal level.

Catering for diversity

It is the teacher's responsibility to ensure that all students progress in their learning. The lessons in these books have deliberately used contexts that are sufficiently broad enough to cater for a wide range of student interests and backgrounds.

The books include a Glossary and Help Boxes to scaffold student learning, while the Challenges promote and extend thinking. Similarly the Revision Challenges units provide an opportunity for students to demonstrate their understanding by answering both closed and open-ended questions and by undertaking various student-centred investigations.

Teaching and learning strategies

Some of the concepts in the books are quite complex and may require a different teaching style. The Papua New Guinea Mathematics Syllabus (2003) promotes the importance of giving students opportunities in mathematics to work cooperatively, discuss and listen to each other's opinions. It is important that students are clear about the problems they are to solve and that they are provided with adequate time to find solutions themselves without being told how to do them. In some cases it may be appropriate for the class to work together on one question rather than work independently on a series of similar questions.

Language

English is the main language of instruction for Upper Primary, although local vernacular can be used to facilitate understanding and reinforce meaning. The Glossary at the back of each book identifies the key words that students need to learn while working through that book. The first few times students meet these key words in the books, teachers should:

- say the word with the class a number of times
- write the word on the board or on a class list or piece of cardboard
- explain the word using real objects, actions or pictures
- demonstrate how to use the word in a simple mathematical sentence
- ask the students to use the word in a simple mathematical sentence
- tell the students to enter the word in their language vocabulary book or the class dictionary.

Materials

The Grade 7 Mathematics books use a range of different materials. Students learn best when they use real materials to help them explore new concepts. Many of the materials that you will need for the program can easily be collected from the students' environment or made from simple items commonly available. Before students use new materials they should be given time to explore them. This will allow them to concentrate on the task when it is finally set instead of wanting to play with the new materials.

In Grade 7 the intention is for students to move progressively away from using real materials when working with numbers to deal directly with symbols. However, with concepts such as fractions, decimals and algebra it may still be preferable to use concrete materials to assist student understanding.

A list of the main materials needed for each topic appears on the double page spread at the start of each topic. It is important that teachers check this list before beginning the topic and ensure that adequate quantities of the materials are available for the students to use. Some of the mathematical equipment listed may need to be purchased or borrowed from other institutions within the community if the school's resources are insufficient.

Materials and equipment should be stored in strong boxes with lids. Students should be told that they are expected to look after materials and to return them to the appropriate box when they have finished using them. Teachers should check all materials regularly and repair or replace any that are damaged.

Assessment

Assessment is the process of finding out what students know. It is important to assess students in order to evaluate what mathematical concepts they understand, how they learn best, what concepts they are ready to learn and how to most appropriately adjust teaching to further support student learning.

Assessment information should be gathered throughout the year using a variety of assessment methods. Written tests and examinations should only form part of the assessment process. Other valuable information can be gathered during everyday teaching by listening to what students say, observing what they do and looking at samples of student work. It is better to collect a small amount of information at many different times throughout the year than to collect a lot of information a few times during the year.

How to evaluate your students

Assessment should be an ongoing process that begins from the first day at school and continues throughout the year. It is important that teachers do not expect students to have mastered all concepts before they have had the necessary experiences and are genuinely able to demonstrate their use of the concepts successfully.

Using open-ended tasks can provide teachers with meaningful information about students that may not be apparent from student responses to closed tasks. The following example shows the difference between an open and closed task about money.

Closed task: What change should be given if K10 is paid for goods that cost K7.35?

Open-ended task: Describe three different ways to give a customer change from K10 for goods that cost K7.35.

There are many possible answers to the open-ended question. A teacher can learn more about what a student understands about money from their responses to the open-ended question than from their one answer to the closed question. Student responses to open-ended questions can show whether they can:

- look at the problem in different ways
- find more than one possible answer
- explain their strategies
- see patterns and make links between their answers

How to record assessment information

Given the variety of assessment strategies that teachers are encouraged to use, it helps if teachers develop a systematic way for recording information about their students. It is important to keep records up to date and ensure that some data is gathered from each student regularly enough to show changes in their skills and knowledge. Assessment should take place at least fortnightly if not weekly.

Some teachers keep student work sample folders that show evidence of the most recent level of student understanding and achievement. Ideally students should make suggestions and discuss with their teacher about which work samples to include in their folder. As a new piece of work shows improved understanding, the older work sample should be removed from the file. Student work sample folders that are kept up to date are a powerful method of monitoring and assessing student progress.

Another strategy is for teachers to keep notes on a class list. Teachers could choose to observe one to two students in each lesson and make comments about what they observe regarding the students' skills, understandings and motivation. The notes could include both positive or negative information and both typical and unusual events. An example of an annotated class list appears below.

Date	Name	Comments
24 March	Elsie	Added all decimals correctly. Used phrase 'zero point eighty-seven'. (May suggest misconception of decimals as whole numbers instead of parts of whole.)
26 March	Fabian	Confused about placement of decimal point when multiplying decimals by decimals.
26 March	Leti	Completed all decimal additions correctly and asked to show others how to do it.

Tests

These may be short answer or longer exercises. An end of unit test by itself provides information too late to be of value in the classroom. Regular short tests during a unit of work provide timely information about student progress and offer a chance for the teacher to change their teaching approach if required in order to improve student outcomes.

Using assessment information

All assessment information should be used primarily to evaluate student performance as a means of refining the teaching approach. There is no value in collecting assessment information if it is not used to inform teaching for the purpose of improving student performance. If the assessment process highlights gaps in a student's knowledge and understandings, it is a signal for the teacher to try a different approach. Assessment helps the teacher to provide better learning opportunities for the student to ensure the desired outcomes are achieved.

Outcomes Map

Strand	Learning Outcome	Topic	Unit
Number and Application	7.1.1	Look At Me	Food and Nutrition
	7.1.2	Look At Me Our Country	Food and Nutrition Water Resources
	7.1.3	Look At Me Our Country	Food and Nutrition Water Resources
	7.1.4	Look At Me	Food and Nutrition
	7.1.5	Look At Me Our Country	Food and Nutrition Which Route?
	7.1.6	Look At Me Our Country	How Do I Compare? Below the Ground
	7.1.7	Our Country	Below the Ground
	7.1.8	Work-wise	Shelf Life
Space and Shape	7.2.1	Look At Me	How Do I Compare?
	7.2.2	Look At Me Our Country	How Do I Compare? Water Resources
	7.2.3	A Day in the Life	Making Things
	7.2.4	Look At Me Our Country	How Do I Compare? Water Resources
	7.2.5	A Day in the Life	Making Things
	7.2.6	Work-wise	Shelf Life
	7.2.7	Look At Me	How Do I Compare?
	7.2.8	Our Country Work-wise	Water Resources Profit Margins
	7.2.9	A Day in the Life	Making Things
	7.2.10	A Day in the Life	Making Things
	7.2.11	Work-wise	Shelf Life
	7.2.12	Our Country A Day in the Life	Which Route? Making Things
	7.2.13	Work-wise A Day in the Life	Shelf Life Time to Play
	7.2.14	Our Country	Which Route?

Strand	Learning Outcome	Topic	Unit
Space and Shape	7.2.15	Our Country	Which Route?
	7.2.16	Our Country	Below the Ground
Measurement	7.3.1	Work-wise	Animal Husbandry
	7.3.2	Look At Me Work-wise	How Do I Compare? Animal Husbandry
	7.3.3	Work-wise	Animal Husbandry
	7.3.4	Work-wise	Animal Husbandry
Chance and Data	7.4.1	Look At Me Work-wise Work-wise	Healthy Living Animal Husbandry Profit Margins
	7.4.2	Look At Me	Healthy Living
	7.4.3	A Day in the Life	Time to Play
	7.4.4	Look At Me Our Country Our Country Work-wise A Day in the Life	How Do I Compare? Which Route? Water Resources Animal Husbandry Making Things
	7.4.5	A Day in the Life Work-wise	Time to Play Profit Margins
	7.4.6	Look At Me Look At Me Our Country Work-wise	How Do I Compare? Food and Nutrition Water Resources Profit Margins
	7.4.7	Work-wise	Profit Margins
Patterns and Algebra	7.5.1	A Day in the Life Work-wise	Tending Gardens Shelf Life
	7.5.2	A Day in the Life A Day in the Life	Tending Gardens Time to Play
	7.5.3	A Day in the Life A Day in the Life	Tending Gardens Time to Play

Glossary

Glossary

abacus (*plural*: abaci) A simple instrument used for counting and calculating.

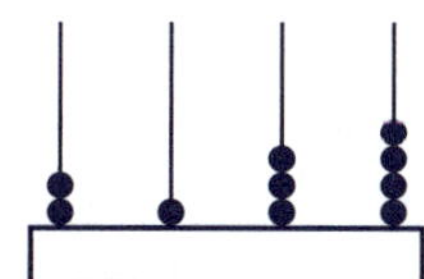

angle An angle is the amount of turn between two arms or rays. The amount of turn is measured in degrees.

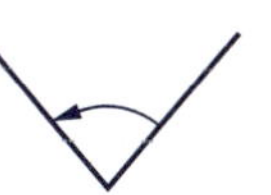

anticlockwise It is the opposite direction to clockwise. It is the opposite direction to the movement of the hands on an analogue clock.

arc A part of a circle or other curved line. You can use a pair of compasses to draw an arc.

area A measure of the total surface of a shape or object. Area is measured in square centimetres (cm^2), square metres (m^2), hectares (ha) or square kilometres (km^2).

average It is also called the *mean* or *arithmetic mean*. It is found by adding a set of scores together and dividing by the number of scores.

axis (*plural*: axes) This is either the vertical or horizontal line that forms the framework of a graph. The horizontal axis is also called the x-axis. The vertical axis is also called the y-axis.

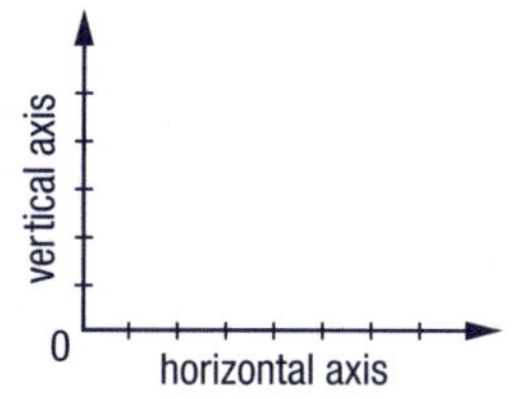

bar graph A graph that uses horizontal or vertical bars to represent information. It is also known as a *column graph*.

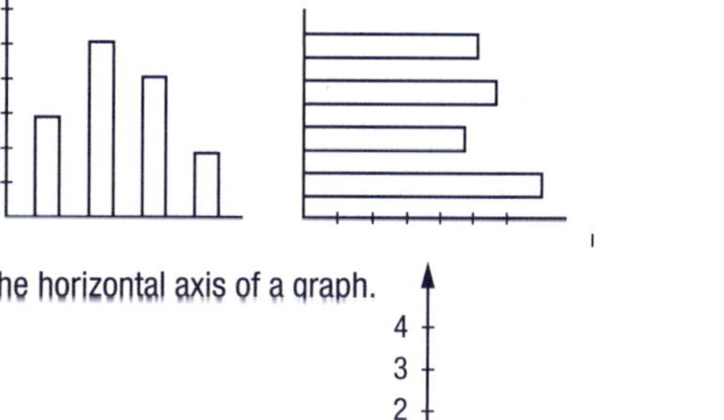

base line The horizontal axis of a graph.

base line

bearing The angle on the ground that either fixes the direction of an object or gives the direction in which something travels. A map and a compass can be used to set a bearing.

bisect To cut or divide into two equal parts. Lines and angles can be bisected.

calculator A machine designed to do mathematical calculations.

capacity The amount of liquid that a container can hold. This is measured in litres (L) and millilitres (mL).

centimetre (cm) A small unit of length. There are 100 centimetres in 1 metre.

centre The middle point of a shape or object.

circumference The distance around the edge of a circle.

classify To arrange into groups according to attributes such as shape, size or colour.

clockwise The direction in which the hands on an analogue clock move.

comparative bar graph A bar graph that combines two or more sets of data on the same graph. A key is used to identify each set of data.

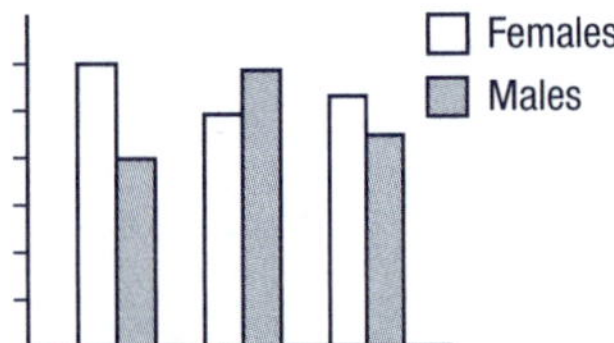

compass There are two different compasses. One is an instrument that is used to find and show direction, and the other is an instrument for drawing circles. The instrument for drawing circles is sometimes called a *pair of compasses*.

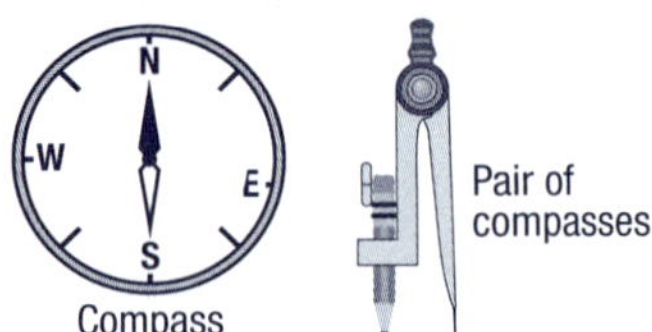

compass points They identify direction. The names of the major compass points are north, south, east and west. This compass shows the direction of the major points:

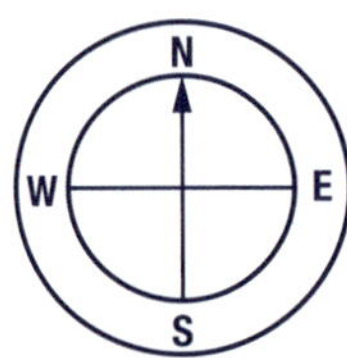

congruent Two shapes are congruent if they have the same size and shape.

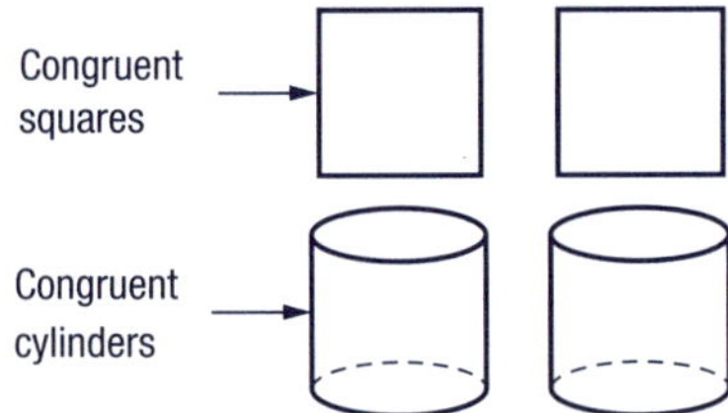

contour line A line on a map joining points of equal altitude or height above the surface of the earth.

coordinates A pair of numbers or letters that show the exact position of something. They are often used on maps and graphs. The horizontal coordinate is given first, for example: the position of the point on this grid is given as B5 or (B,5).

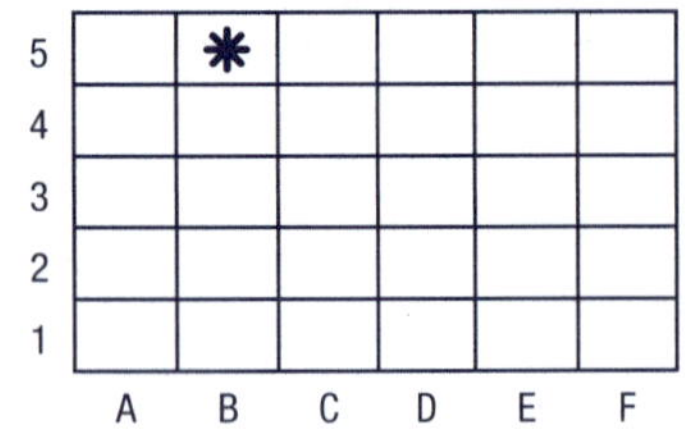

When two numbers are given as coordinates they are usually called *ordered pairs* and are written inside a bracket, for example: (5,2). The horizontal coordinate is still given first. The position of the point on this grid is given as (2,1).

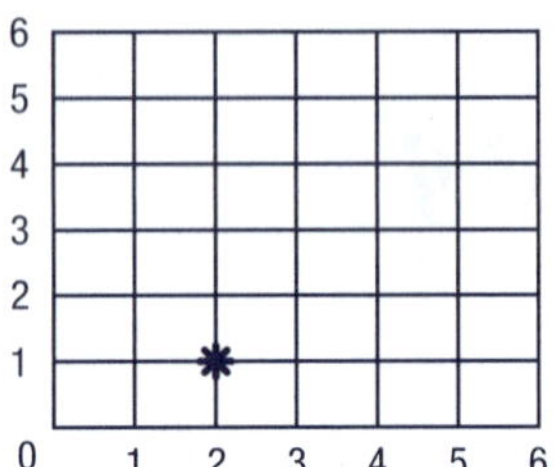

cubic centimetre (cm^3) A unit for measuring volume. It is a cube with all edges of 1 centimetre.

cubit An old measure of length. It was the length from the elbow to the tip of the middle finger of an adult.

data Also known as *statistics*. It is information (numbers or words) that is collected as part of a survey or questionnaire. The data can be shown in a table or on a graph to help us interpret it better.

decimal Fractions such as $\frac{7}{10}$, $\frac{35}{100}$, $\frac{14}{1000}$ and so on, whose denominators (the bottom line of the fraction) are powers of 10, can be written as decimal numbers. Example: $\frac{7}{10} = 0.7$, $\frac{35}{100} = 0.35$, $\frac{14}{1000} = 0.014$. The decimal point shows which digits are whole numbers and which digits are fractions. Those digits to the left of the decimal point are whole numbers (units, tens, hundreds, thousands …) and those to the right are fractions (tenths, hundredths, thousandths …).

decimal number system Commonly called the 'base ten' system, it is a system for writing numbers that uses ten as a base for grouping. Numbers are written using the ten digits 0, 1, 2, 3, 4, 5, 6, 7, 8, 9 and place value. Each place value is a power of ten.

degree A unit for measuring angles. There are 360 degrees in a full turn or circle. The symbol for degree is °.

degree Celsius A common unit for measuring temperature. On the Celsius scale, water freezes at approximately 0 degrees and boils at approximately 100 degrees. The symbol for degree Celsius is °C.

denominator The bottom number written in a fraction. It shows how many equal parts make up the whole. For example: the denominator of ¼ is 4 and it shows that there are four equal parts in the whole. The top number written in a fraction is called the *numerator*.

digit 0, 1, 2, 3, 4, 5, 6, 7, 8 and 9 are digits. Numbers are made up of digits, for example: 256 is a three-digit number made up of 2, 5 and 6.

dimension A measure of size or direction in space. One-dimensional objects, such as lines and curves, have one dimension (length). Two-dimensional objects, such as squares and triangles, have two dimensions (length and width). Three-dimensional objects, such as pyramids and cubes, have three dimensions (length, width and height or depth).

directed number A whole number that has a + or − sign to indicate a positive or negative direction from zero. A directed number is also called an integer. We can show them on a number line.

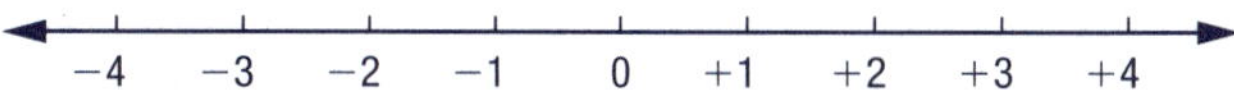

direction Where a moving object is headed or where the position of something is being indicated. Words such as left, right, up, down, above, below, forwards, backwards, north, south, east, west, clockwise, anticlockwise all indicate direction. The + and − signs in front of whole numbers indicate the direction of those numbers from zero.

displacement A method for finding the volume of solids, especially those with an irregular shape. The solid is put into a calibrated measure of water and the amount that the water rises on the scale of the measure is the same as the volume of the solid. For example, if a solid displaces 30 millilitres (mL) of water, then its volume is 30 cubic centimetres (cm^3).

dozen A group of twelve things, for example, one dozen eggs = 12 eggs.

equation A number sentence with an equals sign (=). The numbers on one side of the sign are equal to the numbers on the other side. Examples: $3 \times 4 = 6 \times 2$; $6 + 7 = 13$; $20 + 5 = 30 - 5$ …

equivalent fractions These are fractions that have the same value. For example: $\frac{1}{4}$, $\frac{2}{8}$ and $\frac{4}{16}$ are equivalent fractions. $\frac{1}{4}$, 0.25 and 25% are equivalent fractions.

estimate To estimate is to make an approximate judgement or calculation. An estimate is an approximate judgement or calculation.

factor Any number that you can divide into another number without leaving a remainder, for example: factors of 6 are 1, 2, 3 and 6; factors of 10 are 1, 2, 5 and 10; factors of 3.6 are 1.5 and 2.4 …

fathom A measure of six feet (1.8288 metres) especially used when measuring the depth of water. In the past, a fathom was the distance between your fingertips when your arms were stretched out sideways.

four-quadrant number plane A graph or plane that is divided into four areas or quadrants by two axes crossing at right angles.

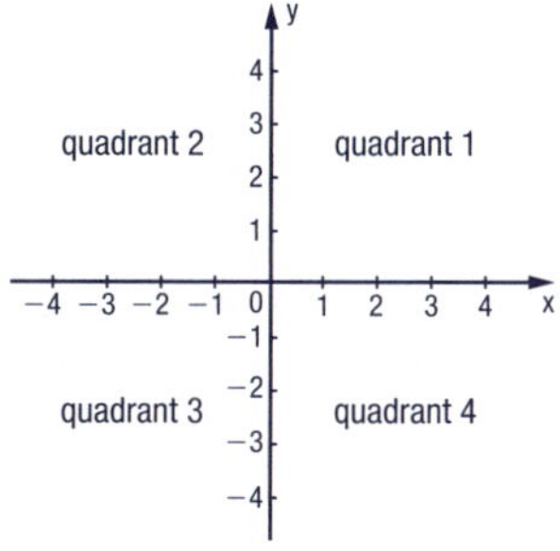

fraction A fraction is part of a group or a whole. It can be written as a common fraction, a decimal fraction, a percentage or a ratio.

1 out of 4 parts = $\frac{1}{4}$

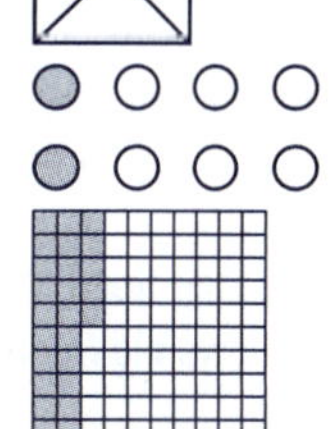

2 out of 8 counters = $\frac{2}{8}$ or $\frac{1}{4}$

25 out of 100 = 0.25 or 25%

Mix 1 part to 4 parts = 1:4

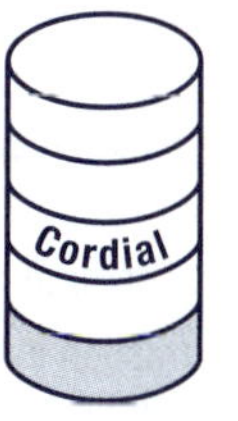

Common fractions are written in the form $\frac{a}{b}$ where a is the numerator and b is the denominator.
A unit fraction is a common fraction which has 1 as its numerator, for example: $\frac{1}{5}$ $\frac{1}{3}$

frequency table A graph or table showing how often an event or quantity occurs.

Numbers thrown on dice		
1	𝍸 III	8
2	𝍸 I	6
3	𝍸 IIII	9
4	IIII	4
5	𝍸 I	6
6	𝍸 II	7

front-end estimation A strategy mainly used for working out an approximate answer to addition and subtraction problems. When adding or subtracting only the first digits are considered when making an estimate. For example:

$$\begin{array}{r} 342 \\ 255 \\ +513 \\ \hline 1000 \end{array}$$

full turn A full turn is a complete revolution or rotation. The angle of a full turn is 360°. An example of a full turn is when a clock hand travels from a number right around the clock face until it reaches the same number again.

gradient A measure of slope or steepness. The gradient of a line is the ratio of the distance risen (the vertical distance) to the horizontal distance travelled.

graduation A graduation is one of the evenly spaced marks on a measuring instrument or scale. For example, the graduations on a thermometer are in degrees; the graduations on a ruler are usually in centimetres and millimetres.

gram (g) A small unit of mass. There are 1000 grams in a kilogram.

graph A drawing or diagram that displays information. There are many different kinds of graphs, for example: bar graphs, pictograms, line graphs and pie charts.

graph title This is the name of the graph and is usually written above the graph. A title makes it clear what the graph is about.

half turn A half turn is half of a complete revolution or rotation. The angle of a half turn is 180°, which is half of 360° – the angle of a full turn.

histogram A type of bar graph that has the bars joined up. It shows the frequency of data in each group.

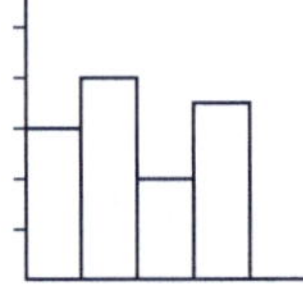

horizontal In the same direction as the horizon. When you are lying down you are horizontal.

hour A unit for measuring time. There are 24 hours in one day.

improper fraction A fraction in which the numerator is greater than its denominator, for example: $\frac{7}{4}$
It is another way of writing a mixed number: $\frac{7}{4} = 1\frac{3}{4}$

integer A whole number or zero. The integers are:

- positive whole numbers: 1, 2, 3, 4, 5, ...
- negative whole numbers: $-1, -2, -3, -4, -5, \ldots$
- zero: 0

intermediate compass points They name the directions that are at 45° angles to the four major compass points. The names of the intermediate compass points are north-east (NE), south-east (SE), north-west (NW) and south-west (SW). This compass shows the direction of the intermediate compass points:

intersect To cross each other. These two lines intersect at Point A:

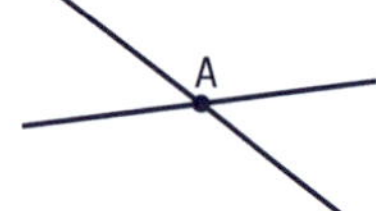

intersection The point at which two or more objects overlap or cross each other. For example:
The point where lines cross.

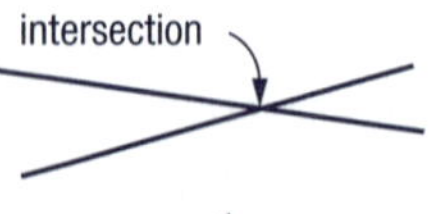

The region where shapes overlap.

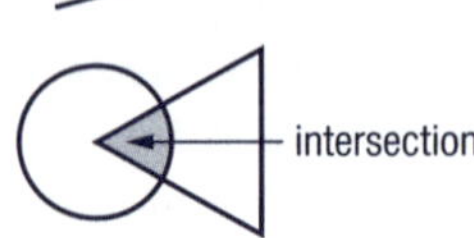

The intersection of two sets is the set of elements that are common to both sets.

key Also known as a *legend*. A list that explains the symbols used on a map, a graph or in a table.

kilogram (kg) A unit of mass. There are 1000 kilograms in a tonne.

kilojoule (kJ) A measure of the energy value of foods.

kilolitre (kL) A unit of volume or capacity used to measure the amount of a liquid. There are 1000 litres in a kilolitre.

kilometre (km) A unit of length or distance. Distances between places are measured in kilometres. There are 1000 metres in a kilometre.

legend Also known as a *key*. A list that explains the symbols used on a map, a graph or in a table.

line A straight or curved length that has only one dimension, length not width. It can have a beginning and end, or it can extend forever.

linear In a straight line. A linear relationship is one that shows the relationship between two quantities on a graph in the form of a straight line.

line graph A graph that shows a trend or relationship. Segments of straight lines connect points that represent certain data.

line segment Part of a line between two given points.

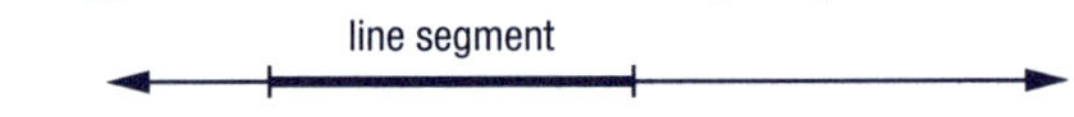

litre (L) A unit of capacity or volume used to measure the amount of a liquid. There are 1000 millilitres (mL) in a litre.

mass The amount of matter in an object. Mass is measured in grams, kilograms and tonnes. Mass is often confused with weight. The mass of an object remains constant from place to place. Weight is the pull of gravity on an object, and can vary from place to place. For example, an astronaut has the same mass on Earth as on the Moon but weighs more on the Earth because the pull of gravity is stronger.

maximum The highest or largest value. For example, the highest temperature for a certain day is referred to as the maximum temperature for that day.

mean It is also called the *arithmetic mean* or *average*. It is found by adding a set of scores together and dividing by the number of scores. For example: A student got marks of 7, 8, 7, 5, 6 and 9 on six different tests. The mean is found by adding the six scores together, $7 + 8 + 7 + 5 + 6 + 9 = 42$, and then dividing by 6 (the number of tests): $42 \div 6 = 7$. So, the mean of the student's marks is 7.

median The middle number in a set of numbers when the numbers are arranged in order. If there is no middle number, the mean of the two middle numbers is taken. In this set of numbers, 12 is the median: 3 5 8 (12) 13 17 26

megalitre (ML) A very large unit for measuring capacity. There are one million litres in a megalitre.

metre (m) A unit of length. There are 100 centimetres in a metre and 1000 metres in a kilometre.

metric system A decimal system of measurement.

metric units Different standard units are used to measure and compare length, mass and volume. In the metric system, the common units of measurement are cm, m, km, g, kg, mL and L.

milligram (mg) A very small unit of mass. There are one thousand milligrams in a gram.

millilitre (mL) A small unit of capacity or volume used to measure the amount of a liquid. There are 1000 millilitres in a litre.

millimetre (mm) A very small unit of length. There are 10 millimetres in a centimetre and 1000 millimetres in a metre.

minimum The lowest or smallest value. For example, the lowest temperature for a certain day is referred to as the minimum temperature for that day.

minute A unit for measuring time. There are 60 minutes in one hour.

mixed number A combination of a whole number and a fraction, for example: $3\frac{1}{4}$ and $5\frac{1}{2}$.

mode The most common or frequently occurring number in a set of numbers. For example: In this set of numbers 5 is the most common: 2 5 8 5 3 2 5 5 9 5

negative number A number less than zero. Negative numbers are always written with a minus sign (−) in front of them, for example: -2, -5.8, $-14\frac{1}{2}$.

network A group of connecting lines. The lines are called *arcs*, and the points where they meet are called *nodes*. The networks below can be traced without lifting the pen from the paper, or going over the same arc twice.

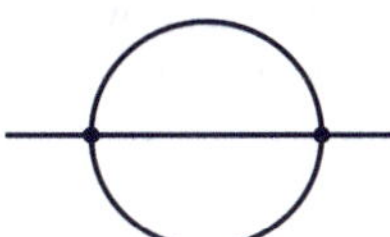

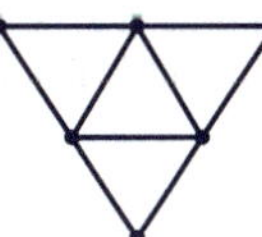

node The point where arcs in a network meet. The network here has five nodes.

number line A line of numbers that shows the correct position of each number in relation to its direction and distance from zero. For example:

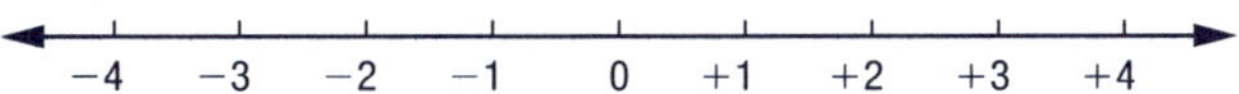

numerator The top number written in a fraction. It shows how many equal parts of a whole have been chosen. For example: the numerator of $\frac{3}{4}$ is 3 and it shows that three equal parts have been chosen out of a possible four equal parts.
The bottom number written in a fraction is called the *denominator*.

ordered pair Two numbers, called *coordinates*, written in a certain order. They are usually written between brackets, for example: (4,2). The order of the numbers is important. The ordered pairs (4,2) and (2,4) are different.

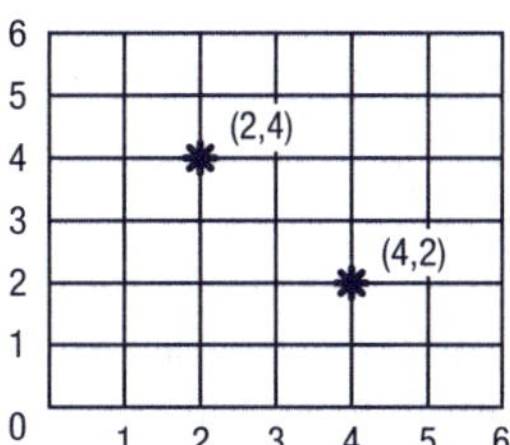

origin The point at which something begins. The point where the axes of a coordinate system cross is called the origin and is marked 0.

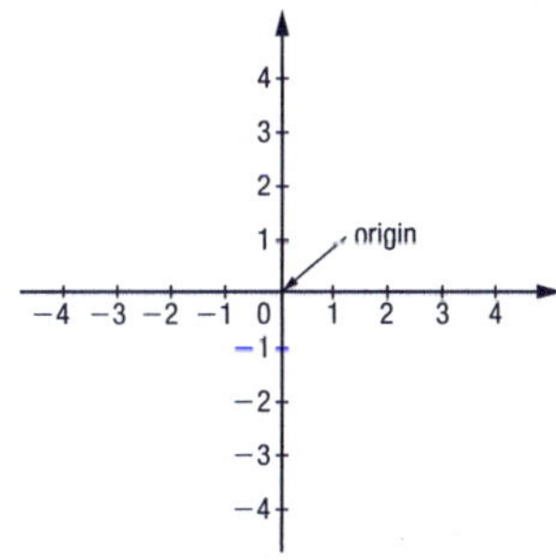

pace The distance between your feet when you take a step. It is measured from heel to heel and is useful for estimating distances.

percentage A fraction expressed in hundredths. The symbol for percentage is %. For example: $\frac{27}{100} = 27\%$; $\frac{1}{2} = \frac{50}{100} = 50\%$.

pictogram A graph where the information is shown in pictures or symbols. Each picture or symbol represents a certain number or amount of something. This is shown by a key somewhere on or near the graph. A pictogram is also known as a *picture graph* or *pictograph*.

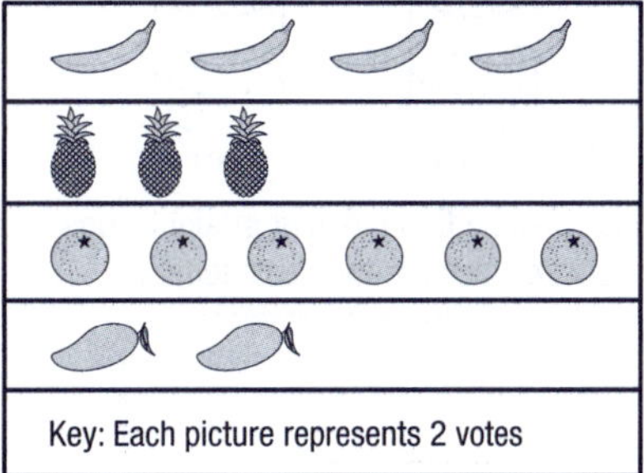

pie chart A graph that is drawn as a circle and shows data as a fraction of the whole, like a pie cut into slices. A pie chart is also known as a *pie graph* or a *circle graph*.

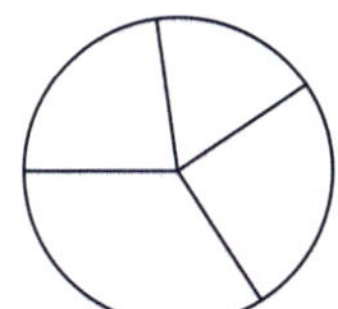

plane shape A shape that can be drawn in one plane. Two-dimensional shapes are plane shapes because they can be drawn on a flat surface.

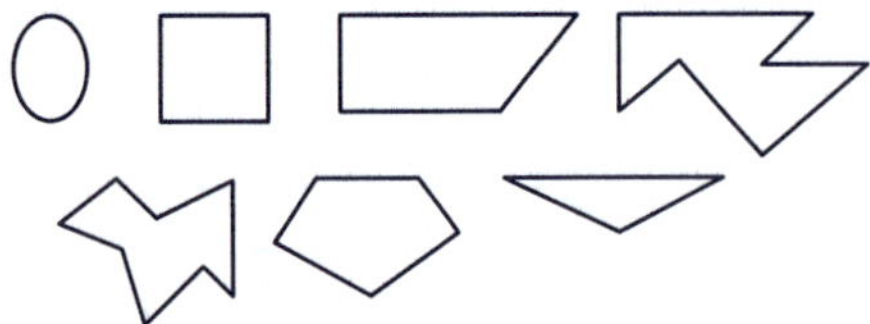

positive number A number greater than zero. Positive numbers are sometimes but not always written with a plus sign (+) in front of them. For example, these numbers are all positive numbers: +3, +7.5, $4\frac{3}{4}$, 24, 15.6.

protractor An instrument for measuring and marking out angles. Its scale is marked in degrees.

quarter turn A quarter turn is quarter of a complete revolution or rotation. The angle of a quarter turn is 90°, which is quarter of 360° – the angle of a full turn.

range Also called *spread*. The difference between the highest and lowest values of a set of data. For example, in the following set of numbers the range is 15 which is the difference between 20, the highest number, and 5, the lowest number: 5, 7, 10, 11, 14, 15, 17, 18, 20.

rate The comparison between two quantities which may be of different things. For example: the rate of water flow from the water tank was 180 litres per hour. This rate compares volume with time.

ratio A comparison of two quantities, measurements or numbers of items. The symbol : is used to express a ratio. For example: when mixing cordial and water in the ratio 1:4, we mix one part cordial to every four parts water.

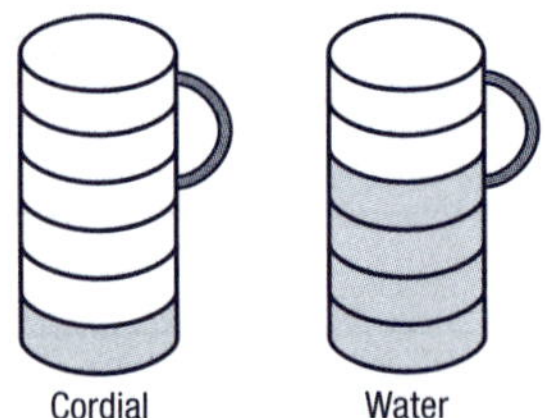

rectangle A plane shape that has four straight sides and four right angles. The opposite sides are the same length.

rounding A strategy used when estimating an answer. Whole numbers, fractions and decimals can be rounded, for example: 278 is nearer to 300 than 200; 6231 is nearer to 6000 than 7000; 4.8 is nearer to 5 than 4; $2\frac{3}{4}$ is nearer to 3 than 2 … Money can be rounded to the nearest kina, for example: K4.15 is nearer to K4 than to K5. When rounding numbers that are halfway between, always go up to the next number, for example: 25 would be rounded to 30; 850 would be rounded to 900; K2.50 would be rounded to K3.

scale A series of evenly spaced marks that we use to measure something. The scale on a ruler is marked in centimetres and millimetres, while the scale on a thermometer is marked in degrees to show the temperature. The scale on a map or plan shows how much the map or plan has been reduced or enlarged, for example: 1 cm = 50 metres.

second A small unit for measuring time. There are 60 seconds in one minute.

sector A region inside a circle that is shaped like a slice taken from the circle by making two cuts. The curved edge of a sector is called an *arc*.

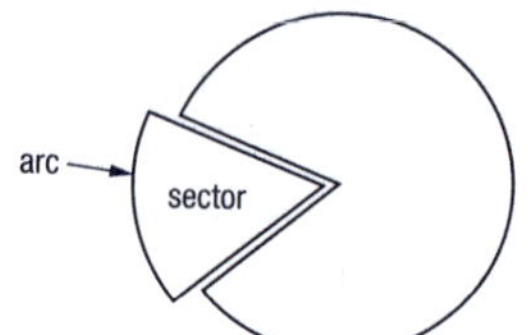

set A group of numbers, shapes or objects with a particular thing in common. For example: The numbers 2, 4, 6, 8, 10, 12, 14, 16 and 18 are the set of even numbers less than 20. The things that belong to a set are called *elements*, so the set described has nine elements.

slope An inclined or slanting position or direction. The amount of slope is called the *gradient*.

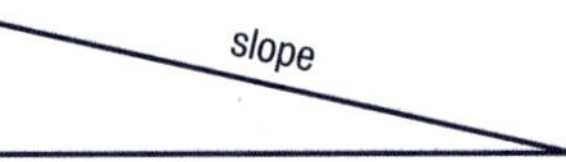

solid A figure with three dimensions. This can be modelled in three forms:

- *solid*: made of solid matter throughout.
- *hollow*: made only of the faces, empty inside.
- *skeleton*: made only of edges.

speed How far and how fast something is moving. For example: A car travelled 60 kilometres in 1 hour, so its speed was 60 km per hour.

spread Also called *range*. The difference between the highest and lowest values of a set of data. For example, in the following set of numbers the spread is 15 which is the difference between 20, the highest number, and 5, the lowest number: 5, 7, 10, 11, 14, 15, 17, 18, 20.

spring balance An instrument for measuring weight.

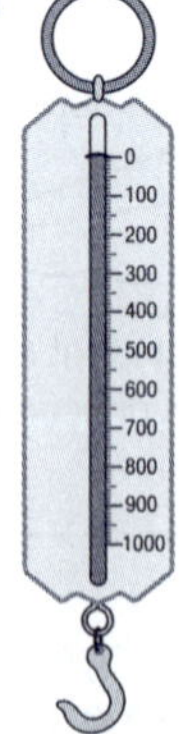

square metre (m^2) A unit for measuring area. The area of a surface is measured by working out how many square metres cover the surface. Other units for measuring area are square centimetres (cm^2), square kilometres (km^2) and hectares (ha).

statistics Also known as *data*. It is information (numbers or words) that is collected as part of a survey or questionnaire. It can be shown in a table or on a graph to help us interpret it better.

stem and leaf plot A graph that displays data in a table. Numbers in the table are broken up into tens digits and ones digits. The tens digits are displayed in the first column and are called the stems. The ones digits are displayed in the second column and are called the leaves. For example, the stem and leaf plot shown here displays the numbers: 5, 8, 12, 15, 15, 16, 19, 21, 26, 27, 34, 39.

Stems (tens)	Leaves (ones)				
0	5	8			
1	2	5	5	6	9
2	1	6	7		
3	4	9			

subset A set within a set. All the elements of one set belong to the other set. For example: If Set A includes all the students in your class and Set B includes all the girls in your class, then Set B is a subset of Set A because all the elements of Set B are also in Set A.

surface The outside of an object. The surface of an object may be flat or curved. The surface of a can is both curved and flat.

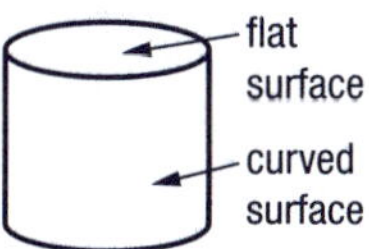

survey A way of collecting data. It can be done by interview, questionnaire or observation.

temperature How hot or cold something is. It is usually measured in degrees Celsius.

three-dimensional (3-D) When something has the three dimensions of length, width and height it is said to be three-dimensional. Solid shapes are three-dimensional.

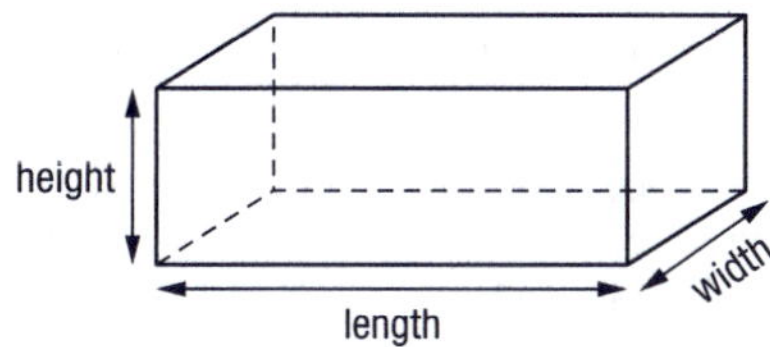

tonne (t) A large unit of mass. 1000 kilograms make a tonne.

tree diagram A diagram that is used to display data. It is so called because it shows the alternatives available by branching out from a starting point. The tree diagram below shows the three-digit numbers that it is possible to make with the digits 2, 3 and 4.

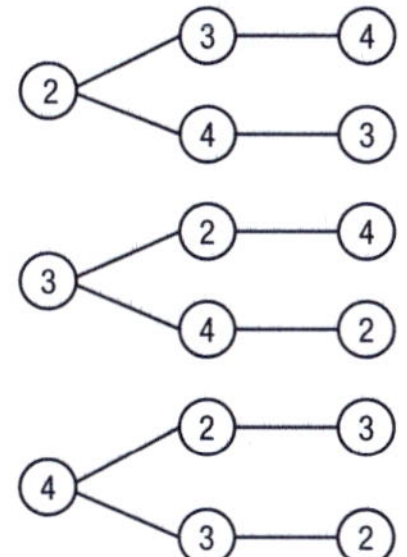

traversable network A network is traversable if it can be traced without lifting your pen or going over any part of it more than once. Are these networks traversable?

two-dimensional (2-D) When something has the two dimensions of length and width, or length and height, it is said to be two-dimensional. Plane shapes are two-dimensional.

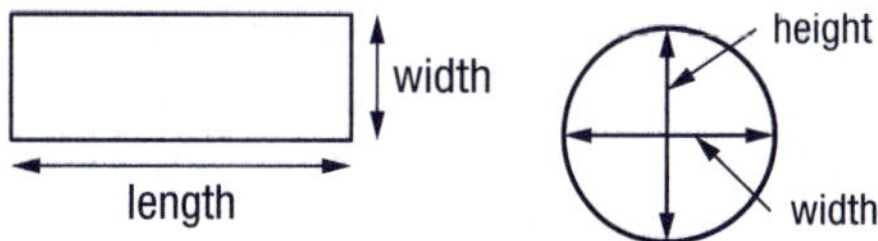

Venn diagram A way of displaying data when items belong in more than one set. The places where the circles intersect show the items that belong in both sets.

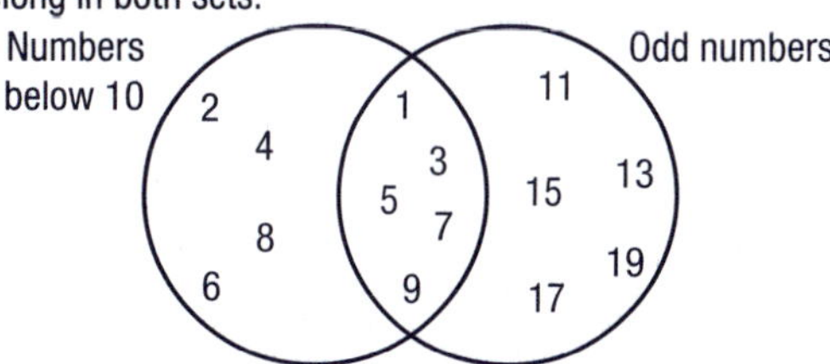

vertex (*plural*: vertices) A point at which two or more lines or edges meet to form an angle or corner.

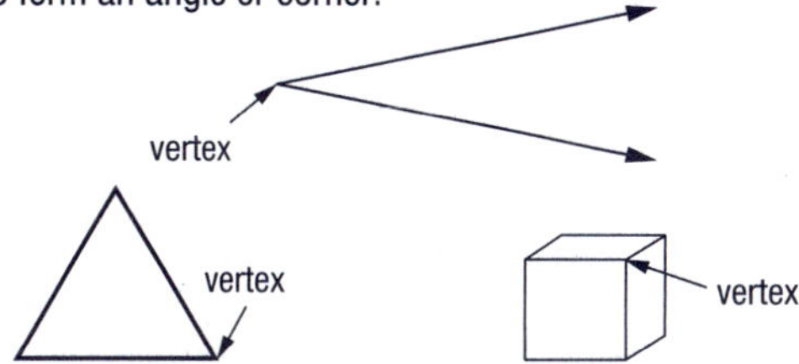

vertical At right angles to the horizon. When you are standing up you are in a vertical position.

vinculum The horizontal line separating the numerator from the denominator. For example $\frac{2}{3}$ ← vinculum

volume The amount of space inside a container or the amount of space a solid occupies. The volume of solids is measured in cubic centimetres (cm^3) or cubic metres (m^3). The volume of liquids is measured in litres and millilitres.

weight The gravitational pull on an object. Weight is measured in grams, kilograms and tonnes. Weight is often confused with mass. The mass of an object remains constant from place to place whereas weight can vary from place to place. For example, an astronaut has the same mass on Earth as on the Moon but weighs more on the Earth because the pull of gravity is stronger.

zero The symbol for nothing or nought (0).

Answers

Topic 1
Learning Unit 1: Food and Nutrition

Lesson 1

1–8 Answers will vary.

Lesson 2

1 a 1100 toea
b 200 toea
c 250 toea
d 1520 toea (anything between 1500 and 1550 is acceptable)
e 150 toea

2 Dairy products seem to be the most expensive. They may be the most expensive because they are not manufactured locally.

3 Vegetables and fruit seem to be the least expensive. This may be because it is easy to grow/cultivate them in Papua New Guinea and there is plenty of agricultural land.

4 a–b Answers will vary.

5 a 262.5t per kg (this would be rounded up to 263t per kg)
b 71.4t per kg (this would be rounded down to 71t per kg)
c 1911t per kg
d 294t per loaf
e 367.5t per 500 g (this would be rounded up to 368t per 500 g)
f 92.4t per 100 g (this would be rounded down to 94t per 100 g)

6 a Sausages and frozen chickens are more expensive per kg in Port Moresby than in Madang *or* Sausages and frozen chickens are cheaper per kg in Madang than in Port Moresby.

6 b Butter and flour are cheaper per kg in Port Moresby than in Madang *or* Butter and flour are more expensive per kg in Madang than in Port Moresby.

7–8 Answers will vary.

Lesson 3

1 1 tomato, 1 small carrot, 1 slice pineapple, 1 boiled egg, 1 slice bread, 110 g sardines, 1 sausage, 1 serve potato chips.

2 Answers will vary. Possible answers are: 1 sausage and 1 boiled egg = 1035 kJ; 1 boiled egg and 1 slice bread and 1 tomato and 1 small carrot and 1 slice pineapple = 990 kJ.

3 Answers will vary.

4 a 4.44 minutes (this may be rounded down to 4 minutes)
b 17.46 minutes (this may be rounded down to 17 minutes)
c 42.12 minutes (this may be rounded down to 42 minutes)
d 25.02 minutes (this may be rounded down to 25 minutes)
e 4.68 minutes (this may be rounded up to 5 minutes)

5 a 22.32 minutes b 2.88 minutes
c 36.72 minutes d 7.68 minutes
e 69.72 minutes or 1 hour and 9.72 minutes

6 Answers will vary.

CHALLENGE: 862.5 kJ; 3290 kJ; 2340 kJ; 2275 kJ; 1216 kJ; 1520 kJ

Lesson 4

1

	kJ per serve	Protein per serve	Fat per serve	Sugar per serve	Sodium per serve
Bread (1 slice)	282.5 kJ	2.3 g	0.4 g	0.6 g	147 mg
Biscuits (4)	209 kJ	1.2 g	1.1 g	0.2 g	66 mg
Difference	73.5 kJ	1.1 g	(0.7 g)	0.4 g	81 mg

2

Energy kJ	Protein	Fat	Sugar	Sodium
1130 kJ	9.2 g	1.6 g	2.4 g	588 mg

3 a 40 g b 32 g c 53 g d 27 g

4

kg of sugar per cake	slices in cake	kg of sugar per slice
0.280 kg	8	0.035 kg
0.360 kg	9	0.040 kg
0.720 kg	16	0.045 kg
1.250 kg	25	0.05 kg
1.5 kg	20	0.075 kg
2.4 kg	50	0.048 kg

Lesson 5

1 a 112.5 g b 337.5 g c 450 g d 562.5 g

2 a 300 mL b 900 mL c 1200 mL d 1600 mL

3 a 900 g b 1350 g c 2700 g d 4500 g

4 a 2:6; $\frac{2}{6}$ b 6:10; $\frac{6}{10}$ c 6:2; $\frac{6}{2}$ d 10:6; $\frac{10}{6}$

5 a 2:9 b 1:3 c 1:2 d 9:15

6 a 1:3 b 1:3 c 1:6 d 1:3 e 1:3
f 1:4 g 1:3 h 3:10 i 3:7 j 7:100

7 a 1:6 b 1:3 c 1:25 d 2:3 e 1:3 f 1:5

8 a 250 g to 1000 g; 1:4 b 250 mL to 2000 mL; 1:8
c 1250 mL to 3000 mL; 5:12 d 750 g to 2500 g; 3:10
e 500 g to 800 g; 5:8 f 2000 g to 50 g; 40:1
g 75 kg to 1000 kg; 3:40 h 1250 mL to 5500 mL; 5:22

9 a 8 b 6 c 2

10 a 4:6 or 2:3 b 3:6 or 1:2

11

Baking tin	Width	Length	Ratio
A	8 cm	12 cm	2:3
B	12 cm	*18 cm*	2:3
C	*16 cm*	24 cm	2:3
D	20 cm	*30 cm*	2:3

12 a K40 to K160 b K60 to K140 c K200 to K300
d K150 to K350 e K30 to K60 f K9 to K81
g K30 to K70 h K20 to K80

13 1:6

14 1320 g of SR flour, 12 tbsp of butter, 6 eggs, 4.5 cups of milk, 4 cups of sugar.

Lesson 6

1 a 10 b 40 c 60 d 60
e 80 f 20 g 70 h 40

2 a 400 b 300 c 900 d 700 e 800 f 300

3 a 2.5 b 59.6 c 47.1 d 1.4 e 92.9 f 31.6

4 a 30 b 180 c 320 d 210 e 4070 f 2120

5 a 200 b 400 c 900 d 1300 e 7900 f 4100

6 a 2.8 b 9.0 c 53.1 d 9.2 e 215.1 f 163.5

7 9700 kJ

8 Anywhere between 8950 kJ and 9049 kJ.

9 Anywhere between 14.5 and 15.49 minutes.

10–12 a–b Answers will vary.

13 a $6 + 4 = 10$ b $8 - 3 = 5$ c $2 \times 6 = 12$ d $9 \div 3 = 3$
e $8 + 2 = 10$ f $4 + 3 = 7$ g $8 \times 2 = 16$ h $6 \div 2 = 3$

14 Possible answers are:
a 25% of 200 = 50, so it will be a bit more.
b 10% of 63 = 6.3, so it will be a bit more.
c 50% of 95 = 47.5, so it will be a bit less.
d 10% of 2468 = 246.8 so it will be a bit less.

15 a 110 b 700 c 100 d 250

16 a 120 b 740 c 140 d 270

17 a 119; 737; 139; 273
b Rounding was closest to the actual answer in all cases.

18

		Front-end estimation	Rounding	Exact answer	Which was closest
a	243 + 549	700	700	792	The same
b	276 + 589	700	900	865	Rounding
c	397 + 207	500	600	604	Rounding
d	513 + 239	700	700	752	The same
e	334 ÷ 89	3.75	3	3.75	Front-end
f	805 ÷ 185	8	4	4.35	Rounding
g	659 × 82	48 000	56 000	54 038	Rounding
h	1127 × 241	200 000	220 000	271 607	Rounding

19 a–h Answers will vary.

20 Front-end estimation: 2680 kJ Exact: 2945 kJ

21 Rounding: 2400 kJ Exact: 2350 kJ

22 Rounding: K17.00 Front-end estimation: K14.00
Exact: K16.93

23

		Front-end estimation	Exact
a	3.9 + 9.7	12	13.6
b	4.03 + 4.32	8	8.35
c	12.35 + 2.7	14	15.05
d	35.035 + 2.57	37	37.605
e	4.2 ÷ 2.1	2	2
f	12.42 ÷ 3.2	4	3.88
g	39.7 × 3.4	90	134.98
h	140.65 × 9.9	900	1392.43

Lesson 7

1 a $\frac{5}{10}$ or $\frac{1}{2}$ b $\frac{3}{10}$ c $\frac{2}{10}$ or $\frac{1}{5}$

2 a

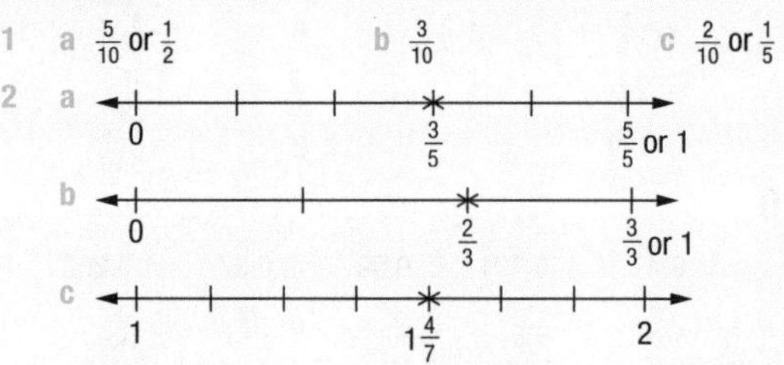

3 a 2 berries b 15 apples c 6 bananas
d 20 oranges e 12 fish f 10 coconuts

4 a $\frac{1}{10}$ b $\frac{1}{4}$ c $\frac{1}{100}$
d $\frac{3}{12}$ or $\frac{1}{4}$ e $\frac{10}{500}$ or $\frac{1}{50}$ f $\frac{2}{12}$ or $\frac{1}{6}$

5 a 500 mL b 750 g c 325 toea
d 8 buns e 400 kg f 40 yams

6 a 2 × 1 litre containers b $\frac{1}{5}$

7 a 12 × 250 mL bottles b $\frac{1}{12}$

8 a 50 × 200 mL bottles b $\frac{1}{50}$

9 Answers will vary.

10 a $\frac{1}{2}$ b $\frac{1}{3}$ c $\frac{1}{4}$ d $\frac{1}{10}$ e $\frac{3}{20}$
f $\frac{1}{4}$ g $\frac{9}{50}$ h $\frac{21}{25}$ i $\frac{3}{500}$ j $\frac{17}{200}$

11 a $\frac{3}{10} > \frac{7}{100}$ b $\frac{3}{7} < \frac{15}{21}$ c $\frac{2}{5} < \frac{15}{20}$ d $\frac{71}{100} > \frac{9}{20}$
e $\frac{13}{15} > \frac{20}{60}$ f $\frac{2}{13} < \frac{15}{39}$ g $\frac{3}{4} = \frac{15}{20}$ h $\frac{4}{5} > \frac{9}{20}$

Lesson 8

1 a $2\frac{5}{6}$ b $4\frac{1}{4}$ c $2\frac{1}{8}$ d $3\frac{3}{10}$

2 a $4\frac{3}{8}$ b $8\frac{1}{10}$ c $6\frac{2}{3}$ d $6\frac{3}{8}$

3 a $6\frac{13}{24}$ b $3\frac{19}{20}$ c $5\frac{9}{10}$ d $6\frac{7}{40}$ e $4\frac{1}{6}$
f $3\frac{3}{40}$ g $6\frac{1}{20}$ h $3\frac{5}{6}$ i $4\frac{33}{40}$ j $5\frac{23}{40}$

4 a $1\frac{1}{12}$ b $1\frac{5}{12}$ c $1\frac{1}{5}$ d $1\frac{2}{3}$

5 a $\frac{5}{8}$ b $3\frac{5}{8}$ c $2\frac{2}{12}$ or $2\frac{1}{6}$ d $3\frac{3}{10}$
e $1\frac{4}{15}$ f $1\frac{8}{10}$ or $1\frac{4}{5}$ g $\frac{1}{30}$ h $1\frac{23}{60}$

6 $2\frac{13}{24}$

7 a $2\frac{1}{4}$ kg
b He bought 72 and his friend gave him 36.

8 a $1\frac{1}{2}$ b $2\frac{5}{8}$ c $3\frac{1}{3}$ d $2\frac{1}{10}$ e $1\frac{1}{2}$ f $\frac{10}{13}$
g $1\frac{1}{6}$ h $6\frac{2}{3}$ i 3 j $2\frac{2}{5}$ k $\frac{6}{7}$ l $\frac{3}{4}$

9 a b $\frac{3}{8}$

10 a $\frac{1}{20}$ b $\frac{9}{40}$ c $\frac{1}{4}$ d $\frac{21}{40}$ e $\frac{6}{91}$
f $\frac{1}{6}$ g $\frac{3}{25}$ h $1\frac{3}{4}$ i $3\frac{21}{50}$ j $8\frac{1}{2}$

CHALLENGE: $\frac{3}{4} \times \frac{1}{3} = \frac{3}{12} = \frac{1}{4}$

11 a $5\frac{1}{2}$ b $9\frac{1}{3}$ c $13\frac{1}{3}$

12 a $1\frac{2}{25}$ b $1\frac{7}{8}$ c $1\frac{5}{16}$ d 6 e $\frac{8}{9}$
f $6\frac{2}{5}$ g $2\frac{2}{9}$ h $\frac{2}{3}$ i $\frac{5}{12}$ j $\frac{15}{16}$

13 a $\frac{50}{69}$ b $1\frac{43}{65}$ c $3\frac{4}{15}$ d 2 e $5\frac{19}{20}$

Lesson 9

1 a 0.3 b 0.47 c 0.701 d 0.09 e 0.37 f 0.263

2 a $\frac{78}{100}$ b $\frac{617}{1000}$ c $\frac{3}{100}$ d $\frac{5}{10}$ e $\frac{29}{1000}$ f $\frac{12}{100}$

3 a $1\frac{3}{10}$ b $3\frac{3}{50}$ c $7\frac{1}{250}$ d $3\frac{1}{5}$ e $4\frac{1}{20}$ f $10\frac{3}{4}$
g $2\frac{37}{100}$ h $4\frac{1}{8}$ i $7\frac{13}{50}$ j $2\frac{3}{40}$ k $5\frac{1}{250}$ l $22\frac{51}{200}$

4 a 0.1 kg b 0.2 kg c 0.375 kg
d 0.6 kg e 1.5 kg f 1.1 kg

5 a 0.429 b 0.444 c 0.833 d 0.667 e 0.583 f 0.417

6 a 3.6 kg b 3.35 kg

7 3.75 kg

8 a 30% b 59% c 25% d 63%
e 5% f 40% g 130%

9 a 0.97 and $\frac{97}{100}$ b 0.39 and $\frac{39}{100}$ c 0.05 and $\frac{5}{100}$ or $\frac{1}{20}$
d 0.18 and $\frac{18}{100}$ or $\frac{9}{50}$ e 1.0 and $\frac{100}{100}$ or $\frac{1}{1}$ f 0.25 and $\frac{25}{100}$ or $\frac{1}{4}$
g 1.15 and $1\frac{15}{100}$ or $1\frac{3}{20}$

10

Fraction			$\frac{1}{4}$		$\frac{4}{5}$			$1\frac{3}{10}$	
Decimal	0.3	0.12		0.125		1.25	2.6		3.0
Percentage	30%	12%	25%	$12\frac{1}{2}$%	80%	125%	260%	130%	300%

11 3.170 kg and $3\frac{17}{100}$ kg

12 Answers will vary.

CHALLENGE: Lillian spends 15% on food. Lillian spends: K37.50 on food; K125 on bills; K62.50 on entertainment; K25 on clothes.

Lesson 10

1 a grams b kilograms c kilograms
d grams or kilograms e tonnes

2 Answers will vary.

3 a 500 g b 1750 g c 1250 g d 2125 g e 2400 g f 667 g

4 a–e Answers will vary.

5 $1\frac{4}{15}$ cups 6 $2\frac{3}{8}$ kg or 2375 g

7 1.61 kg 8 Answers will vary.

9 There is enough to make 7 full loaves and half another loaf.

10 35% remains.

11 a K1.44 b K1.51 c Shop A

Topic 1
Learning Unit 2: Healthy Living

Lesson 1

1–7 Answers will vary.

Lesson 2

1 a 11
b Cricket
c Rosa and Kila

2

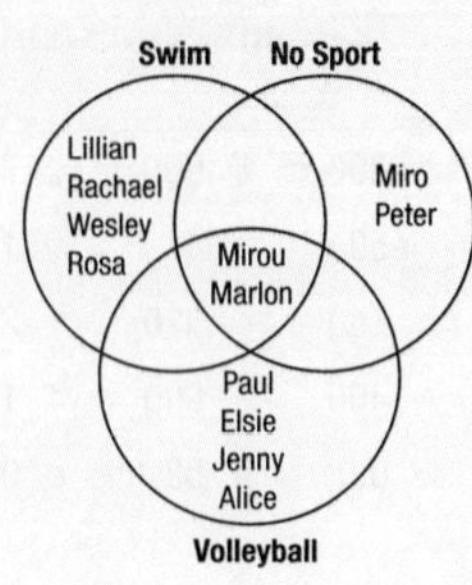

3 Answers will vary.

4 a–b Answers will vary.

5 a Even numbers more than 14 and less than 20
b Odd numbers below 20

6 Answers will vary.

7 a b

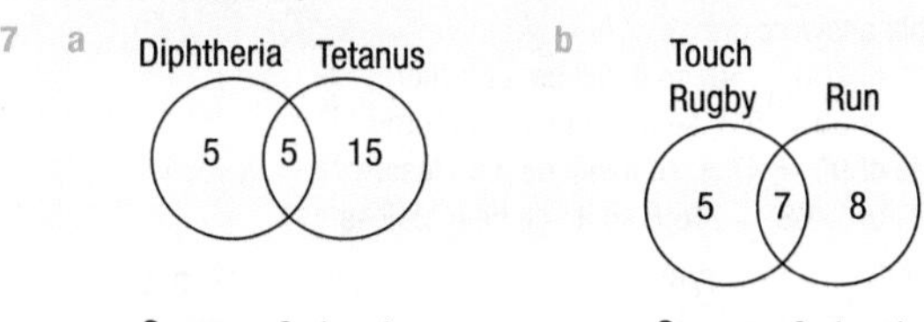

c

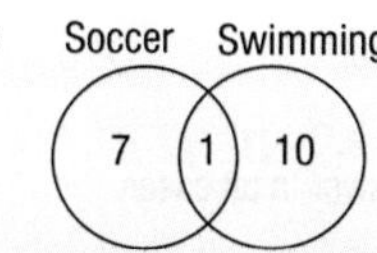

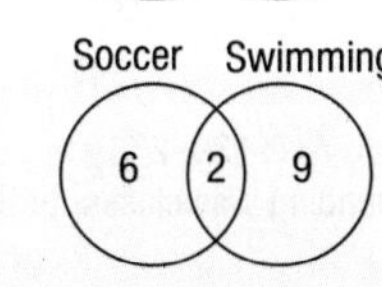

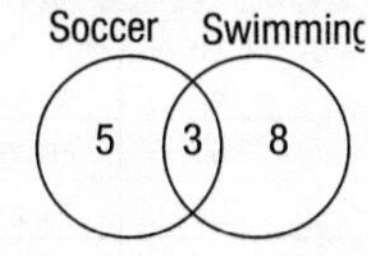

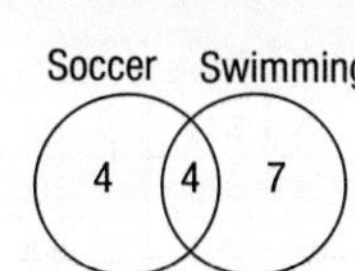

Lesson 3

1 a line graph b bar graph (comparative)
c pie/circle graph d bar graph (stacked)

2–7 Answers will vary.

Lesson 4

1 a 10 minutes b Keti c 5 minutes d Mirou

2 a–b Answers will vary.

3 a 4 b Pineapple c Mango d 1

4 a

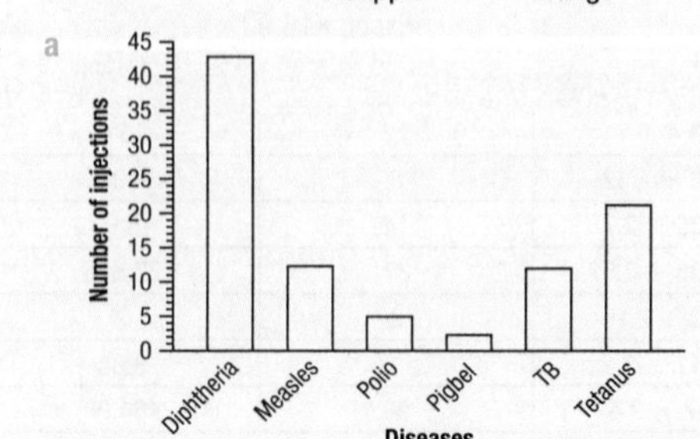

b Answers will vary.
c x-axis: Diseases y-axis: Number of injections
d Answers will vary.

Lesson 5

1 a Catholic b Baptist c 23%
d Does not practise any of the religions surveyed
e 1%

2–3 Answers will vary.

4 a 15–64 b 900

5 Answers will vary. 6 a–c Answers will vary.

7 a 0–20 b Females
c The 0–20 age group is the largest, followed by the 21–30, the 31–40, the 41–50 and the 50+. The number of people in each age group decreases steadily as the population ages.

8 a 5–10 b 5–10 and 11–15 c 70

9 a–b Teacher to check.

Lesson 6

1 a Life expectancy of males has increased.
b 60
c Answers will vary but should be more than the life expectancies shown on the graph.
d Answers will vary but could include: better medical facilities and better nutrition.

2 a The UV radiation levels in Port Moresby month by month.
b x-axis shows the months of the year; y-axis shows UV levels.
c November
d June and July
e Answers will vary.

3 a Answers will vary but possible answers are: doctors, nurses, parents, nutritionists.
b Healthy
c Between 72 and 90 kg
d Below 40 kg

4 a The graph is about the diseases people died from in Papua New Guinea in 1992.
b 400 c 200 d Pneumonia

5 a Yes b–c Second graph
d Different scales have been used on the two graphs.
e Yes. The scale of the x-axis in the 1st graph is not suitable for the data.

6 a–b Answers will vary.

Lesson 7

1 a The graph describes the percentage of a person's time spent on various activities during one 24-hour day.
b 8% of the day is spent playing sport.
c 17%
d 6 hours (25% of 24 hours)
e 90° (right angle)

2 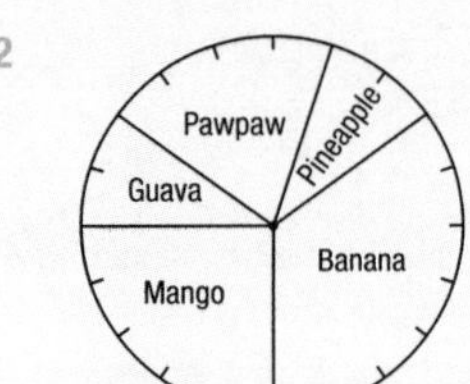

3

Infant immunisation program			
Vaccine	**Number of babies immunised**	**Fraction of total**	**Angle of sector in pie graph**
Diphtheria	70	$\frac{70}{200}$	$\frac{70}{200}$ of 360° = 126°
Measles	50	$\frac{50}{200}$	$\frac{50}{200}$ of 360° = 90°
Polio	30	$\frac{30}{200}$	$\frac{30}{200}$ of 360° = 54°
TB	15	$\frac{15}{200}$	$\frac{15}{200}$ of 360° = 27°
Pigbel	35	$\frac{35}{200}$	$\frac{35}{200}$ of 360° = 63°
TOTAL	**200**	$\frac{200}{200}$	**$\frac{200}{200}$ of 360° = 360°**

CHALLENGE: A pie graph might look similar to this:
Written questions will vary.

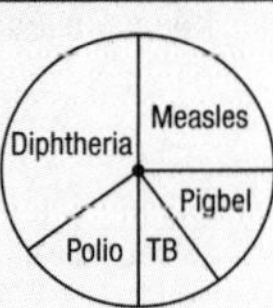

Lesson 8

1 a 4 b 3 c 4 d 2
e P Lokei or A Ure or T Ure

2 Answers will vary.

3 a Protein, fat, carbohydrates
b Saturated, polyunsaturated, monounsaturated
c Sugars and Other

4 a 12 different types:

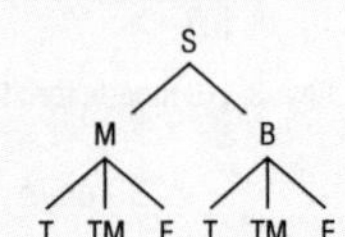

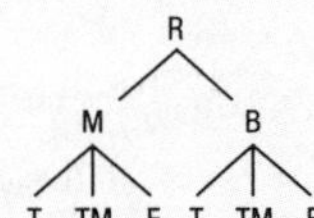

b 8 choices:

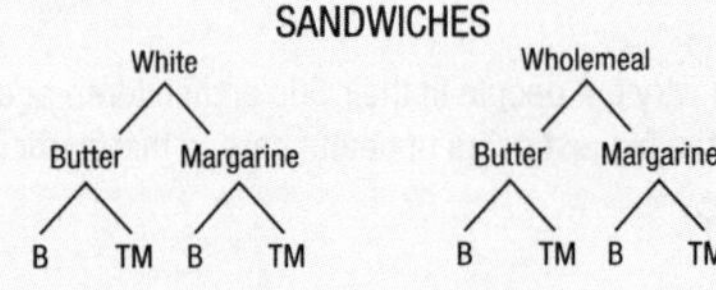

c 24 choices:

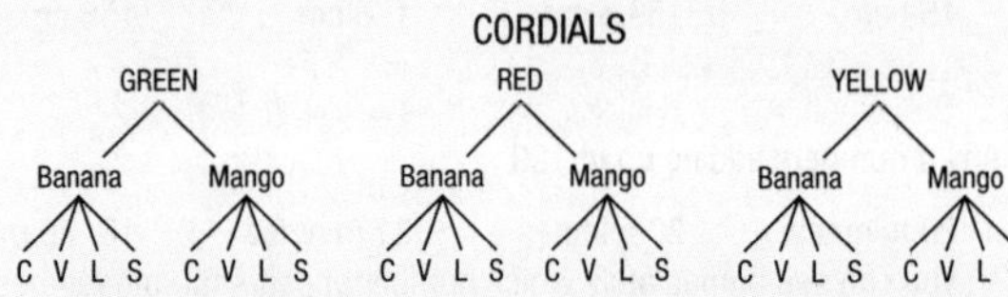

d Multiply the number of choices in line 1 by the number of choices in line 2 by the number of choices in line 3 by the number of choices in line 4. For Question 4c this would be 3 × 2 × 4, which is 24 choices.

5 a 9 combinations:

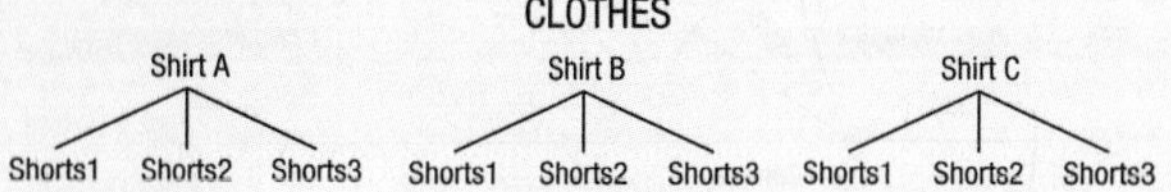

b $3 \times 3 = 9$

6 cricket or soccer, cricket or basketball
cricket or tennis, cricket or rugby
soccer or basketball, soccer or tennis
soccer or rugby, basketball or tennis
basketball or rugby, tennis or rugby

7 a Answers will vary.
b Use a tree diagram, calculate or write out a list.

Lesson 9

1 a

Stem	Leaf
0	5, 6
1	0, 2, 5, 6, 9

Key: 1 | 0 stands for 10

b

Stem	Leaf
0	6, 9
1	1, 3, 6
2	4, 7, 9
3	1
5	3

Key: 2 | 4 stands for 24

2 a

Stem	Leaf
0	3, 7, 9
3	2
4	1, 7
6	2, 4

Key: 6 | 2 stands for 62

b

Stem	Leaf
1	0, 3, 4, 4, 5
2	2
3	2

Key: 2 | 2 stands for 22

c

Stem	Leaf
0	0, 2, 3, 6, 8, 9
1	0, 1, 1, 5, 5

Key: 0 | 0 stands for 0

d

Stem	Leaf
1	2, 5, 8
2	3, 5, 9
3	9
4	5

Key: 3 | 9 stands for 39

3 a 25 b 0 days c 33 days

4 a 1, 1, 1, 2, 3, 11, 11, 19, 24, 28, 32, 52, 53, 53, 56, 56, 58, 59, 62
b 19
c 1 and 62
d Answers will vary but people in their 50s and children aged 3 and below were the biggest users of health care in that particular week.

5 Answers will vary.

Lesson 10

1 a 159 cm b 159.5 cm c 156 cm d 156 cm – 163 cm

2 24

3 Any 4 numbers adding up to 100

4 a 20 minutes b 20 minutes c 21 minutes d 18–23 minutes
e You can see immediately which number appears the most times.

5 Answers will vary but must total 15 kg.

6 Answers will vary.

7 a The percentage of salary that is spent on different food types
b The y-axis shows the percentage of salary.
c Mean would be drawn through the 4.7% line
Median would be drawn through 4% line
Mode would be drawn through 3% and 5% line

8–9 a–c Answers will vary.

Topic 1
Learning Unit 3: How Do I Compare?

Lesson 1

1 a–e Answers will vary. 2 Teacher to check.

3–4 Answers will vary.

Lesson 2

1 a 50 cm b 73 kg c 190°C
d 13 minutes past 5 e 35°C f 85 kph
g 2.8 kg h 300 mL

2 a 450 ml b 38°C c 36 mm d 500 g e 15 mL f 50°

3 a weight – kilograms and grams
b height – metres and centimetres
c time – seconds
d length/distance – metres
e capacity – litres and millilitres
f length – metres
g angle – degrees
h length/thickness – millimetres
i temperature – degrees Celsius

4 Teacher to check. 5 Answers will vary.

Lesson 3

1 Answers will vary.

2 Suitable units are:
a grams b kilograms c kilograms d grams
e kilograms f milligrams g tonnes h grams
Estimates of weight for specified items will vary.

3–4 Answers will vary.

5 It is possible to use different devices for weighing but the following are probably the most appropriate:
a spring balance b bathroom scales c pan balance
d pan balance e kitchen scales

6 a 3 kg b 8000 kg c 1.25 g d 4.75 t e 57 000 g
f 2500 g g 4750 kg h 2500 g i 2 250 000 g

7 Answers will vary.

8 About 2.5 bags

9 a Teacher to observe
b Answers will vary but should be somewhere between 10 and 13 students with bags.

Lesson 4

1 Suitable units are:
a metres b centimetres
c metres or kilometres d metres or kilometres
e millimetres
Estimates of the length or distance of the specified items will vary.

2 Estimates will vary.
a 4.9 cm or 49 mm b 1.3 cm or 13 mm c 7.9 cm or 79 mm
d 10.7 cm or 107 mm e 5.8 cm or 58 mm

3 a–d Answers will vary.

4 a–g Answers will vary.

5 a Tuan 94 cm Pete 85 cm Sera 82 cm
Lily 87.7 cm Luana 91.3 cm
b Tuan, Luana, Lily, Pete, Sera

6 a 37 mm, 390 mm, 30 mm; 390 mm, 37 mm, 30 mm
b 82 mm, 9.2 mm, 8.5 mm; 82 mm, 9.2 mm, 8.5 mm
c 300 m, 480 m, 40 m; 480 m, 300 m, 40 m
d 800 m, 9.5 m, 30 m; 800 m, 30 m, 9.5 m

7 Teacher to check.

8–9 Answers will vary.

10 a 311.7 m b 2702 cm c 118.4 m

11 1.25 km

12 2.3 km or 2300 m

13 3 laps

14 Daniel jumped 2.67 m. Bill jumped 2.95 m.

Lesson 5

1–3 Answers will vary.

4 Answers will vary, but a major reason is: you could not run 20 km at the same speed as you run 10 metres.

5 a 7200 beats per hour b 261 skips in 3 minutes
c 4 hours d 150 claps in one minute

6 Heartbeat if the person stays at rest all the time.

7 a 3 minutes b 12 minutes c 2 km

Lesson 6

Area of each shaded square:

1 a a = 4 cm^2; b = 5 cm^2; c = 9 cm^2; d = 7 cm^2; e = 16 cm^2
b a = 8 cm; b = 12 cm; c = 20 cm; d = 14 cm; e = 20 cm

2 a–d Answers will vary. 3 a–b Answers will vary.

4–5 Answers will vary.

6

Length	Width	Area
7 m	8 m	56 m^2
12 cm	5 cm	60 cm^2
8 m	4 m	32 m^2
4 m	12 m	48 m^2
10 mm	14 mm	140 mm^2
Answers will vary, but must multiply by width to equal 24	Answers will vary, but must multiply by length to equal 24	24 m^2

7–8 Answers will vary.

9 a 32 m^2 b 43 m^2 c 28 m^2 d 93 m^2 e 106 m^2

10 Answers will vary.

Lesson 7

1 a 8 L b 750 mL c 50 mL

2 a 1000 mL b 1000 L c 1750 mL
d 250 mL e 400 L f 800 L

3

mL	500	1000	1500	2500	3000	3500	5000
L	0.5	1	1.5	2.5	3	3.5	5

4 a 1.25 L b 0.375 L c 1.8 L d 3.2 L e 4.95 L
f 1.08 L g 2.866 L h 1.008 L i 1.005 L

5 a 2500 mL b 8250 mL c 300 mL d 1356 mL e 4050 mL
f 2550 mL g 12 050 mL h 650 mL i 10 010 mL

6 a Estimates will vary. A standard cup holds 250 mL of liquid.
b Answers will vary.

7 Assuming a standard measuring cup holds 250 mL of liquid:
a 2275 mL Day 1; 2100 mL Day 2; 1800 mL Day 3
b Day 1
c 2058.3 mL
d Her daily fluid intake is above the recommended daily intake.

8 a–d Answers will vary. 9 a–b Answers will vary.

10 a 6.25 mugs, so 6 full mugs b 2 mugs each
c 100 mL d 3.5 mugs each

Lesson 8

1 a 50 cm^3 b 500 cm^3 c 2000 cm^3 d 2500 cm^3

2 a 40 mL b 75 mL c 1.5 L d 10 L

3 a–c Answers will vary.

4 a–g Answers will vary.

5 Answers will vary.

6 a–b Answers will vary.

CHALLENGE: 2 500 000 L
2500 m^3
Possible dimensions include:
50 m × 25 m × 2 m
50 m × 50 m × 1 m
25 m × 100 m × 1 m

Topic 1
Learning Unit 4: Revision Challenges

Lesson 1

1 a $\frac{7}{20}$ b $\frac{7}{50}$ c $\frac{1}{4}$ d $\frac{47}{100}$
e $\frac{2}{25}$ f $\frac{1}{10}$ g $\frac{99}{100}$

2 a 0.03; 3% b 0.17; 17% c 0.8; 80% d 0.6; 60%
e 0.6; 60% f 0.29; 29% g 0.56; 56%

3 a K9 b 200 L c 95 g
d 180 m e 1125 cm f K2050

4 a 8 minutes b 24 minutes c 1 hour 16 minutes

5 a 13.5 g b 4.8 slices

6 a 1:3 b 1:4 c 1:4 d 1:5

e $\frac{1}{4}$ f $\frac{1}{5}$ g 600 mL h 1.2 L

i 600 mL cordial and 2400 mL water

7 a 1:3 b $\frac{1}{3}$ c 1:3 d $\frac{1}{6}$ e 1:2

f 7:3 g 9:16 h 4:3 i 1:1 and $\frac{1}{4}$

8 a 24 cm × 40 cm b 9 cm × 15 cm c 15 cm × 25 cm

Lesson 2

1

3.86 ↓

3.8 3.85 3.9

2 a 200 b 2400 c 3700 d 32 500 e 10 700 f 900

3 a 30 b 80 c 7000 d 300 e 3160 f 4100

4 a 4.4 b 2.8 c 18.1 d 35.3 e 2.1 f 10.3

5 Any number from 350 to 449

6 Any number from 75 to 84

7 Any number from 3.15 to 3.24

8 a–d Answers will vary.

9 1211; 445; 12 282; 19.9

10 a 24 b 12 c 20 d 39 e 15

11 7 more cakes

12 a $1\frac{1}{9}$ b $1\frac{1}{10}$ c $\frac{23}{40}$ d $4\frac{7}{12}$ e $4\frac{19}{70}$

13 a $\frac{3}{10}$ b $\frac{5}{12}$ c $\frac{3}{8}$ d $\frac{5}{12}$ e $\frac{13}{40}$

14 a $\frac{3}{32}$ b $\frac{1}{4}$ c $6\frac{1}{24}$ d $\frac{2}{15}$ e $\frac{9}{14}$

15 a $6\frac{2}{3}$ b $1\frac{4}{5}$ c $\frac{8}{9}$ d $4\frac{1}{4}$ e $2\frac{9}{10}$

16 a 1.5 kg b 1.8 kg c 2.85 kg d 16.5 kg e 0.92 kg f 0.325 kg

17 $3\frac{5}{6}$

18 $2\frac{1}{8}$ L

Lesson 3

1–2 Answers will vary.

3 Mean 34.5; Median 30.5; Mode 26

4

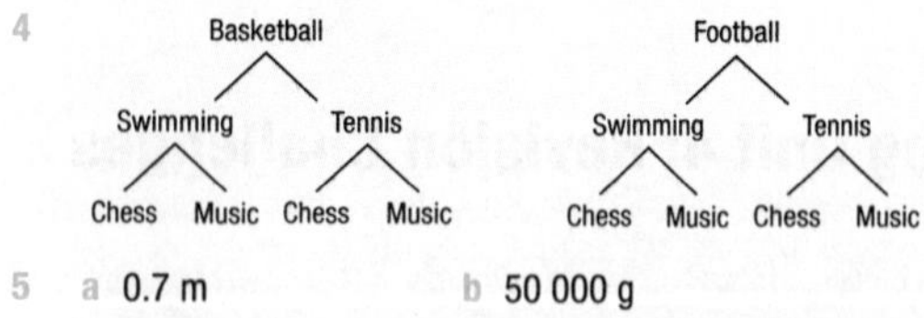

5 a 0.7 m b 50 000 g

c 60 cm^2 d 300 mL (small vase) or 5 L (large vase)

6 a 5 km b 1500 mL c 92 000 g d 4200 m e 25 000 mg

f 2500 cm g 4.32 g h 800 g i 2.75 L

7 10 000 cm^3

8 100 mL

9 a 1 km b 20 km

Topic 2
Learning Unit 1: Which Route?

Lesson 1

1 Answers will vary.

2 1 cm = 13 000 cm (1 cm = 130 metres)

3 1.105 km

4–5 Answers will vary.

Lesson 2

1 a Western Province (Fly) b Central

c Bougainville (North Solomons) d Morobe

e Northern (Oro Province)

2 Answers will vary.

3 a (3.5, 1.25) b (3, 3.5) c (6.5, 4)

d (1.75, 4.5) e (3.5, 2.5)

4 Answers will vary.

5 a V b B c N d H

6 A = (1,4); C = (3,4); D = (4,1); E = (6,2); F = (5, 4.5); G = (1,1); J = (8, 1.5); K = (2,5); Z = (2,1)

7 a–b Answers will vary.

8 Possible coordinates are: (1, 5); (0.5, 5); (0.5, 6); (0, 5); (0, 6); (1.5, 6)

9 a (4.75, 1.25) b (3, 2.5) c (4.5, 2.5)

d (3, 1.5) e (2, 3.5) f (1, 3.5) or (1, 4)

10–11 Answers will vary.

Lesson 3

1 a Quarter turn to the left b 45° turn to the left

c 135° turn to the right d Half turn to the left (or right)

e 135° turn to the right

2 Child A – the dog; Child B – Child D/ the lighthouse; Child C – Child E/ the shop; Child D – the wreck; Child E – the sea; Child F – the fire

3 a the lighthouse b the fire

c The playground d the shop

4 a square

5–6 Answers will vary.

CHALLENGE: Answers will vary.

7 a 45° b 100° c 30° d 65° e 120° f 145°

8 a 30° b 50° c 130° d 25° e 80° f 100°

Lesson 4

1 a W b O c T d R

e P f S g M h X

2 Answers will vary.

3 a 45° b 135° c 225° d 315°

4 K = N45°W; L = N80°E; M = S40°E; O = S70°W

5 M = N15°W; P = N60°E; Q = S30°E; R = S80°W

6 a NE b NW c NW d SE

Lesson 5

1 a ∠ MNO, ∠ OMN, ∠ O b ∠ XYZ, ∠ ZYX, ∠ Y
c ∠ HEB, ∠ BEH, ∠ E d ∠ DEF, ∠ FED, ∠ E
e ∠ LNP, ∠ PNL, ∠ N f ∠ JKL, ∠ LKJ, ∠ K

2 Possible diagrams:

a

b P Q R

3 • = ∠ CAF or ∠ FAC
* = ∠ EAF or ∠ FAE
■ = ∠ EDF or ∠ FDE
▲ = ∠ ECF or ∠ FCE

4–5 Teacher to check.

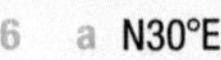

6 a N30°E b S60°E c N90°E d S45°E e N65°E

7 a 45° b 30° c

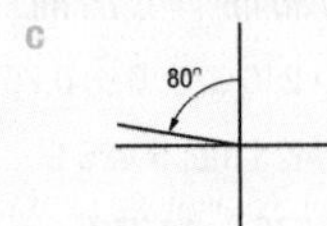

d

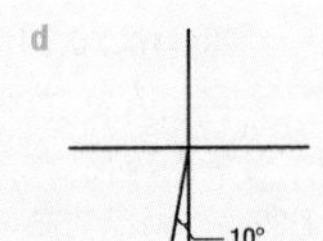

e

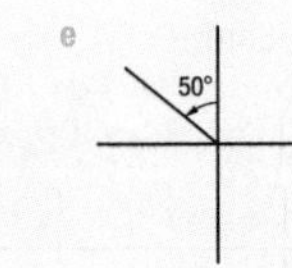

8 a Done b NW c 90° d NW e 180°

9 ∠ AZY and ∠ YZB (∠ YZA and ∠ BZY)

10 ∠ XZU and ∠ UZB (∠ UZX and ∠BZU)

11 45°

12 30°

13 ∠ QGP (∠ PGQ) or ∠ PGL (∠ LPG)

14 60°

15 90°

16 Teacher to check. The angles measure 30°. The sum of their degrees is 60°.

17 Answers will vary.

Lesson 6

1 a 100 cm or 1 metre b 650 metres

2 a between 500 and 550 metres b between 500 and 550 metres
c about 1500 metres (1.5 km)

3 a–c Answers will vary

4 a 0.5 cm b 3240 metres (3.24 km)

5 a 648 km b 540 km

6 a 4 cm b 22 cm c 49 cm d 1 cm e 4 cm
f 18 cm g 21 cm h 41 cm i 39 cm

7 a

cm on map	1	3	10	0.5	12.5
actual km	5	15	50	2.5	62.5

b

cm on map	1	5	7	2.5	23.5
actual km	3	15	21	7.5	70.5

8 a 2 cm b 3.25 cm c 5.75 cm

a	4 km	1 km
b	6.5 km	1.625 km
c	11.5 km	2.875 km

9–10 a–b Answers will vary.

11 Answers will vary.

Lesson 7

1 a 1: 200 000 b 1:10
c 1:100 d 1:50 000
e 2:100 000 (1:50 000) f 2:2 000 000 (1:1 000 000)
g 2:6 (1:3) h 3:10
i 4:200 (1:50)

2 a 40 m b 100 m c 120 m d 12 m e 36 m
f 380 m g 10 m h 24 m i 82 m

3 a Every 1 cm on the map represents 2 500 000 cm (250 km) in actual distance.
b 1 cm = 250 km; 10 cm = 2 500 km; 30 cm = 7 500 km

4

Width	Length	Ratio
2	3	2:3
8	12	2:3
14	21	2:3
24	36	2:3

5 Teacher to check.

Lesson 8

1 a between 300 m and 450 m b between 200 m and 300 m
c B and E d Town D
e H to G is relatively flat whereas B to D is much steeper and would require more effort.

2 a 2000 m b 600 m c 2200 m d 700 m

3 a Mt Bellamy b Templeton's Crossing c Kokoda Gap

4 a b

c The island has a gentle slope up to a hill and in the distance is a tall peak.

Lesson 9

1–4 Answers will vary.

5 a, c and e are traversable.

6–7 Answers will vary.

8 a and b are traversable.

Topic 2
Learning Unit 2: Water Resources

Lesson 1

1–5 Answers will vary.

Lesson 2

1 a
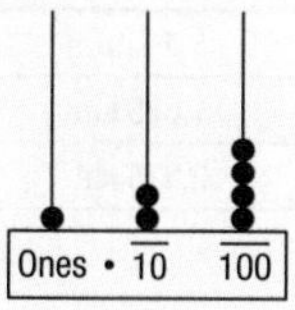

b
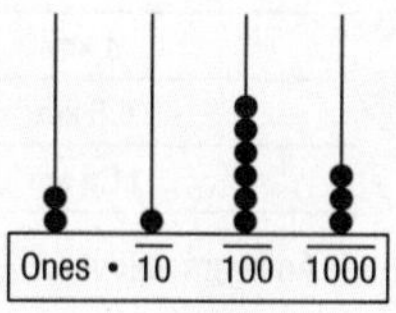

c
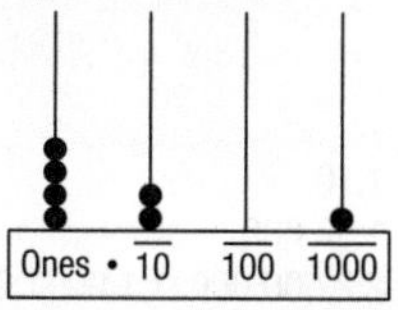

d
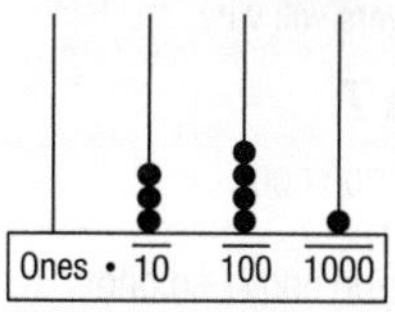

e
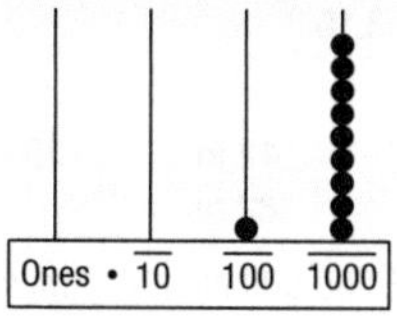

2 a 6 ones b 6 hundredths c 6 tenths
d 6 thousandths e 6 tenths

3 a 7.63 b 5.07 c 0.29 d 0.967 e 2.902
f 0.809 g 10.099 h 0.989 i 17.024

4 a $2\frac{7}{10}$ b $4\frac{36}{100}$ $(4\frac{2}{9})$ c $\frac{427}{1000}$ d $2\frac{9}{100}$
e $5\frac{3}{1000}$ f $2\frac{49}{1000}$ g $\frac{49}{1000}$ h $8\frac{209}{1000}$

5 0.28 < 0.9; 2.333 > 2.04; 1.012 < 1.120; 0.55 < 0.6; 3.078 < 4.1; 3.27 > 3.072; 1.005 < 1.101

6 Answers will vary somewhere between 5348.9 and 5600.7.

7 62.4 km wider

8 Answers will vary between 1150 km and 1249 km.

9 a February b May
c 15.5 cm; 15.25 cm; 14.9 cm; 13.8 cm; 11.4 cm; 11 cm

10 Answers will vary but must be between 3 L and 3.5 L of water.

11 a 2.75 b 3.3 c 5.7 d 10.354 e 1.9 f 14.363

12 a–f Answers will vary.

13 a 3.05 b 1.55 c 2.35 d 34.075
e 21.53 f 0.65 g 0.775 h 198.05

Lesson 3

1
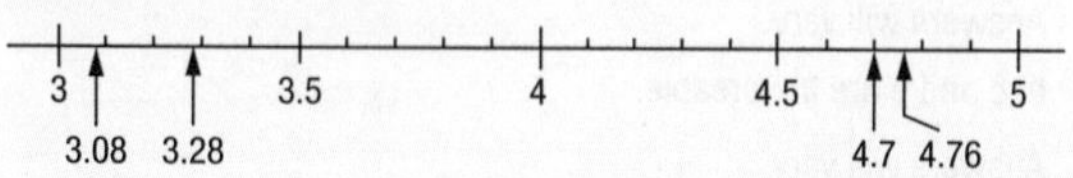

2
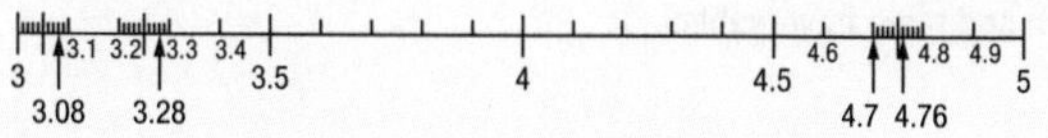

3

Rounded to three decimal places	Rounded to whole number	Rounded to one decimal place	Rounded to two decimal places
2.547	3	2.5	2.55
4.866	5	4.9	4.87
6.871	7	6.9	6.87
7.903	8	7.9	7.90

CHALLENGE: Answers will vary between 10 500 km and 11 499 km.

4 a 0 b 133 c 20 d 17
e 3 f 100 g 23 h 20

5 a 3.8; 2.4; 27.1; 10.0; 4.9; 140.3

6 Answers will vary.

CHALLENGE: Answers will vary.

Lesson 4

1 200 mL = 0.2 mL; 600 mL = 0.6 mL; 2 L = 2.0 L;
750 mL = 0.75 mL; 10 L = 10.0 L; 375 mL = 0.375 mL

2 0.2, 0.375, 0.6, 0.75, 2.0, 10.0

3 5425 mL; 5.425 L

4 4575 L; 4.575 L

5 a 1.2 L b Sam c 300 mL; 0.3 L

6

0.08 less	0.2 less	Starting number	0.25 more	0.75 more
6.92	6.8	7	7.25	7.75
5.22	5.1	5.3	5.55	6.05
9.67	9.55	9.75	10	10.5
9.98	9.86	10.06	10.31	10.81
30.26	30.14	30.34	30.59	31.09

7 Answers will vary.

8 2.85 m 9 1.2 m 10 0.5 m 11 153.2 L

12 Estimates will vary.
a 39.83 b 44.02 c 6.81
d 191.67 e 67.991 f 12.043

13 a–d Answers will vary.

14 Different answers are possible.

Lesson 5

1 a $209.3 \times 10 = 2093$ b $1.456 \times 10 = 14.56$
c $52.65 \times 100 = 5265$ d $29.723 \times 1000 = 29\,723$

2 a 35 L b 350 L

3 38 mm 4 54 325 L 5 175 L 6 188.1 mm

7 a 648.2 b 691.2 c 4462 d 1236
e 1080.54 f 549.06 g 2151.25 h 811.55

8 a K27 b K121.50

CHALLENGE: K20.25; K33.75; K270.
Descriptions of the method used will vary.

9 17.5 L 10 K18.60

11 a 246.4 b 642.6 c 963.7 d 221.03
e 426.02 f 165.242 g 43.995 h 4266.92

12 3.938 L 13 K50.19

14 a 17.15 b 49.68 c 5.78 d 7.59 e 918 f 240.01

Lesson 6

1 a 1.486 b 9.7637 c 0.27456 d 0.3408 e 0.3904

2 1.275 m 3 5.5073 L 4 0.98 toea

CHALLENGE: 350 mL; 0.35 L

5 1.938 kg 6 15.81 m

7 a 2.188 b 4.851 c 1.758 d 0.021
e 2.813 f 0.458 g 0.781 h 0.368

CHALLENGE: 118 bottles; 0; 300 mL

8 1.688 L

9 K2.47

10 Estimates will vary.
a 29.87 b 9.97 c 17.59 d 7.61
e 5.30 f 1.38 g 1.43 h 10.05

11 17 families 12 15 lengths

13 Estimates will vary.
a 2 b 674.30 c 3.24 d 5.12
e 11.79 f 9.41 g 91.21 h 0.42

Lesson 7

1 November 2 September 3 October 4 Pomio

5 Goroka 6 1110 mm 7 995 mm

8 Goroka; June; July; August – the months with lowest rainfalls

CHALLENGE: Goroka – 161.08 mm: Kokoda – 297.92 mm; Pomio – 471 mm; Rabaul – 167.67 mm; Wewak – 178.5 mm

9

Rainfall in millimetres						
Monday	Tuesday	Wednesday	Thursday	Friday	Saturday	Sunday
9	12	13	8	6	10	5

10 3 mm 11 63 mm 12 9 mm

13 Goroka with 254 mm in February. Calculation: 63 mm per week in Feb × 4 = 252 mm

Lesson 8

1 Estimates will vary.
a 3.24 m^2 b 35.88 m^2 c 53.07 m^2 d 36.05 m^2 e 303.8 m^2

2 13.12 m^2

3 12.15 m^2; 7.7 m^2; 3.19 m^2; 21.38 m^2; 76.19 m^2

CHALLENGE: Two possible dimensions are: 4.5 m × 2.8 m; 3.5 m × 3.6 m. There are other possible dimensions.

4 31.28 m^2

5 a

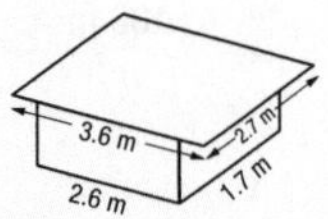

b Estimates will vary; 9.72 m^2

6 Answers will vary.

Lesson 9

1 Divide area by width to calculate length.

2

Length	Width	Area
5 m	9 m	45 m^2
3 m	12 m	36 m^2
3.5 m	4 m	14 m^2
2 m	5.6 m	11.2 m^2
12.4 m	5 m	62 m^2
6.2 m	8.2 m	50.84 m^2

CHALLENGE: The length and width of the roof can be any two numbers that multiply together to give 32. For example: 8 m × 4 m; 2 m × 16 m; 6.4 m × 5 m.

3 a 8 m b 6.12 m c 6.2 m d 10.1 m e 7.02 m

4 a 30 m b 5 m c 21 m d 7.94 m e 20.1 m

5 The dimensions of Peter's roof can be any two numbers that multiply together to give 500. For example: 20 m × 25 m; 10 m × 50 m; 5 m × 100 m.

6 Teacher to check answers. Possible answers are: 5 m × 2 m; 4 m × 2.63 m; 3.5 m × 3.14 m.

7 a 2251.33 L each
b Answers will vary but must total 6754 L.
c Large tank holds 3377 L and smaller tanks hold 1688.5 L each.

Lesson 10

1 2 L; 3 L; 4 L; 9 L; 10 L; 4 m^2 ; 5 m^2 ; 360 L; Answers will vary.

2 2 m^2

3 a 2136 L b 312 L c 3288 L d 6084 L

4 12 000 L

5 If the roof was located in Lae, 55 200 L could be caught. This is more than 4 times the amount that could be caught in Port Moresby.

6 13.52 m^2

7 135.2 L

8 7.4 mm

CHALLENGE: Answers will vary

Topic 2
Learning Unit 3: Below the Ground

Lesson 1

1 Answers will vary.

2 Mt. Kare

3–5 Answers will vary.

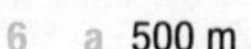

6 a 500 m b 900 m c 100 m

7 a −250 m b −175 m

8 Answers will vary.

Lesson 2

1 a 5 km under the speed limit
b K45 loss
c gaining 3 kg in weight
d running 2 km less than yesterday
e 30 m below ground
f 10 seconds before blast-off
g losing by 20 m
h going down 9 steps
i a profit of K40
j 14°C below freezing
k add 12
l subtract/take away/minus 9

2 Answers will vary.

3 a +3 b −7°C c +K25 d +5 kg
e −2 m f +3 points g −5 seconds h −K10
i +30°C j −5 floors k −K20 l −3

4 a-b

+12, +11, +10, +9, +8, +7, +6, +5, +4, +3, +2, +1, 0, −1, −2, −3, −4, −5, −6, −7, −8, −9, −10, −11, −12

c −9; −7; −6; −2; 0; +3; +8; 9; 11

5 +4 > −2, −4 > −5, −1 < +1, −4 < +3, −5 < −4,
+2 > −3, −2 < 0

6 a −45 < +18 b +75 > +45 c −28 > −35 d −62 < 0
e +27 > −39 f −98 < −25 g −14 < +53 h +10 > −1

7 a −8, −3, +4, +5, 7
b −5, −3, −2, +1, +4
c −5, −4, +7, 10, +12
d −7, −1, +2, +6, 9
e −43, −34, +5, +34, 43
f −28, −15, 0, 14, +31
g −51, −45, 27, +53, +67
h −34, −8, 0, +10, +23
i −97, −45, 15, +28, 78

Lesson 3

1 13 m from the bottom of the cliff

2 Answers will vary.

3 a −3 b 2 c 10 d −9

4 a −7 + +3 (Start at −7 and go up 3)
b 5 + −3 (Start at 5 and go down 3)
c −5 + −4 (Start at −5 and go down 4)
d 2 + 5 (Start at 2 and go up 5)

5 a 3 b 3 c −5 d −10 e 11
f 7 g 1 h 0 i −6 j −4

6 Teacher to check number lines.
a Start at −4. Go right 6. Finish at +2.
b Start at +5. Go left 3. Finish at +2.
c Start at −8. Go left 2. Finish at −10.
d Start at +4. Go right 6. Finish at +10.

7 74 m

8 250 m

9 400 m

10 Teacher to check number line.
a −5 + +8 = +3
b +8 + −9 = −1
c −3 + +5 + −7 = −5
d +7 + +3 + −5 = +5
e −1 + +5 + −9 + +3 = −2

Lesson 4

1 Teacher to check number lines and descriptions.
a −4 b +1 c +3 d −2 e +4

2 Teacher to check number lines.
a +1 b +3 c +2 d +3

3 Answers will vary.

4 The direction of movement on a vertical number line is upwards.

5 a 19°C b 36°C c 12°C d 2°C e −3°C

6 a 12°C b 17°C c 21°C d 25°C e 26°C

7 25°C at night, 48°C during the day.

CHALLENGE: Add 28°C to the village temperature.

Lesson 5

1 a Site B b Site D c 245 m d 642 m e 842 m

2 a Level 15
b He did not go above ground level.
c Level 10
d Up 10 levels

3 a −12°C b −12°C at noon c −4°C at midnight
d −10°C e 8°C below zero

4 Stories will vary.

5 a K178 b K178 c K213

6

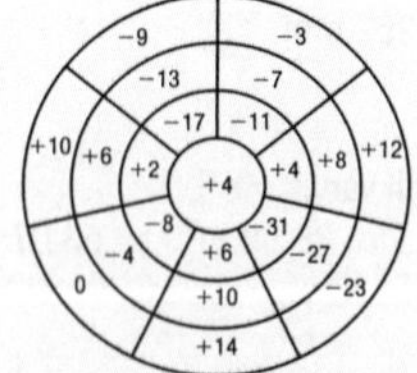

CHALLENGE: 0, +1, −4, −1, +2, −3, −2

Lesson 6

1 a (+2, +3) b (−3, −3) c (+4, −2) d (−1, +2)

2

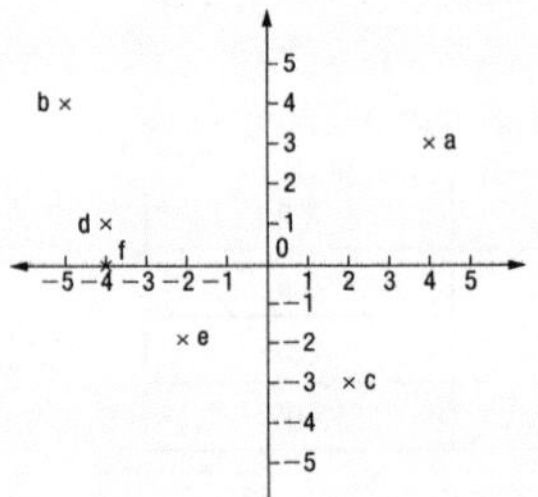

3 a = (+1, −2) b = (+2, +1) c = (0, +2) d = (+1, +4)
e = (−3, +1) f = (−4, −2) g = (−1, −4) h = (+3, −1)

4

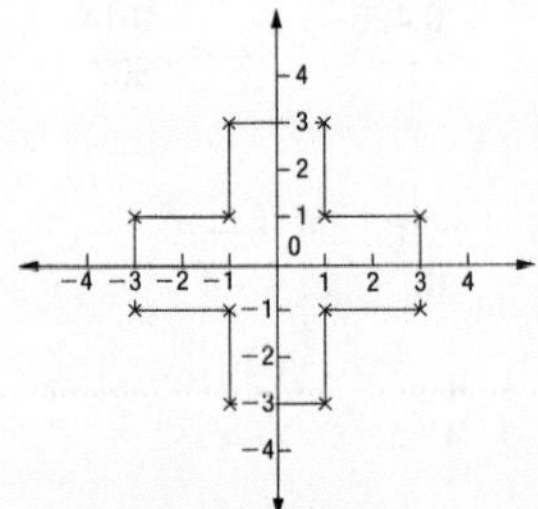

5 Shape A: (+2, +3) (−1, +3) (−3, +1) (0, +1)

Shape B: (+3, +2) (−3, −1) (+1, −2)

6 Shape A is a parallelogram; Shape B is a triangle.

7 Answers will vary.

CHALLENGE: Answers will vary.

Lesson 7

1 a

Depth	3 m	6 m	9 m	12 m	15 m
Seconds	5	10	15	20	25

b

Days off	4	8	12	16	20
Days worked	14	28	42	56	70

c

Grams	1	10	100	500	1000
Price	K18	K180	K1800	K9000	K18 000

2 a

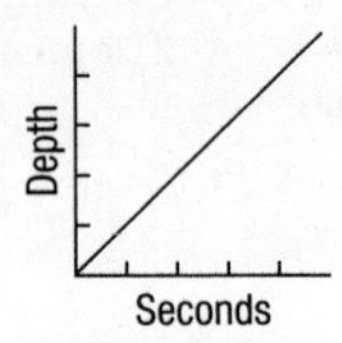

Days off

Days worked

b Answers will vary.

3

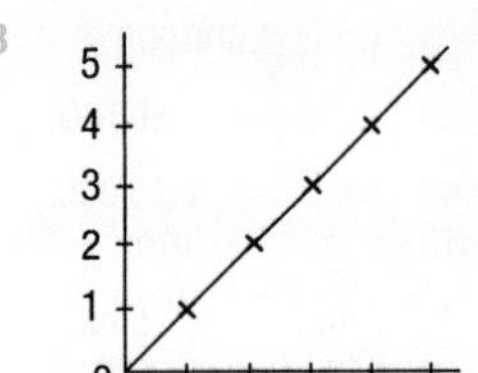

4 a E b D c E, A, B, C, D d E

5 A = positive; B = negative; C = zero; D = positive; E = negative; F = positive; G = negative; H = positive; I = zero

6 A = zero; B = negative; C = positive; D = positive; E = negative

Lesson 8

1 a 1:6, $\frac{1}{6}$ b 1:3, $\frac{1}{3}$ c 1:3, $\frac{1}{3}$ d 2:1, $\frac{2}{1}$
e 1:3, $\frac{1}{3}$ f 100:1, $\frac{100}{1}$ g 300:1, $\frac{300}{1}$

2 a metres per minute b grams per tonne
c litres per hour d heartbeats per second
e kilometres per minute f Kina per hour
g kilojoules per hour

3 a 2:3 b positive

4 a 1:2 b positive

CHALLENGE: Answers will vary.

5 a–c Teacher to check.

6 a A = negative gradient; B = zero gradient; C = positive gradient
b Graph C
c Graph A: with a negative gradient, the miner is getting closer to the main tunnel as time goes on.
Graph B: with a zero gradient, the miner remains the same distance from the main tunnel as time goes on.
d Answers will vary.

7 a The slope would have a negative gradient.
b

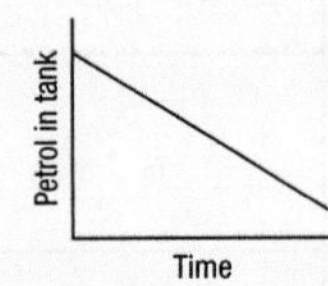

c The more time the truck is driven the less petrol remains in the tank.

Topic 2
Learning Unit 4: Revision Challenges

Lesson 1

1 a (4, 2) b (5, 2.5) c (1.5, 2) d (3.5, 4.5)
e (3, 4) f (2.5, 4) g (6, 4.5) h (4, 3.5)

2 Teacher to check. The shape drawn will look like this:

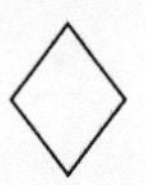

3 a N10°W b N45°E c S30°W d S70°E

4 a 25 km b 9 cm c 1:500 000

5 a 1:100 000 b 3:200 c 1:20 d 1:500 000

6 a 1:20 b 1:9 c 1:4 d 1:20 000
e 1:4 f 1:500 g 1:10 h 30:1

7 a 15 cm b 6 m c 30 mm d 6 kg e 75 mL
f 9 mm g 18 L h 12 min i 150 m j 30°C

8

Scale	Ratio	Fraction
1 cm = 3 m	1:300	$\frac{1}{300}$
2 cm = 5 cm	2:5	$\frac{2}{5}$
3 cm = 5 km	3:500 000	$\frac{3}{500\,000}$
4 cm = 100 km	1:2 500 000	$\frac{1}{2\,500\,000}$

9 a 100 m b C

c 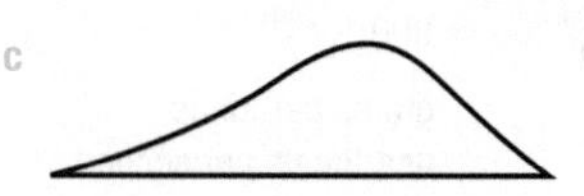 d 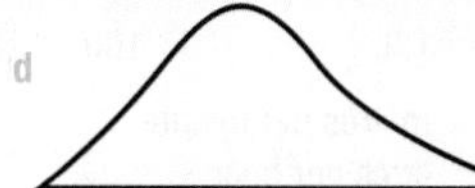

10 Yes, it is possible to travel from one village to another without visiting each village more than once.

Lesson 2

1

Abacus	Decimal	Fraction
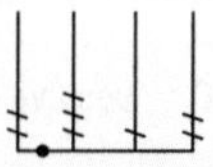	2.312	$2 + \frac{3}{10} + \frac{1}{100} + \frac{2}{1000}$
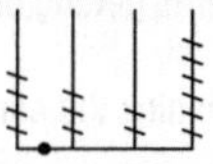	4.326	$4 + \frac{3}{10} + \frac{2}{100} + \frac{6}{1000}$
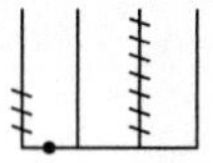	3.07	$3 + \frac{0}{10} + \frac{7}{100}$
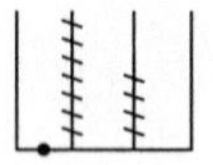	0.74	$\frac{7}{10} + \frac{4}{100}$
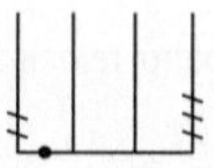	2.003	$2 + \frac{0}{10} + \frac{0}{100} + \frac{3}{1000}$

2 4.326; 3.07; 2.312; 2.003; 0.74

3 a 2.4 b 4.2 c 3.3 d 4.1
e 2.0 f 1.9 g 0.1 h 0

4 a 3 b 4 c 6 d 2
e 2 f 3 g 4 h 3

5 6.72, 6.27, 2.76, 2.67, 2.6, 2.07, 2.06, 0.68, 0.673, 0.67

6 a–c Answers will vary.

7

0.5 less	Number	0.5 more
2.2	2.7	3.2
−0.2	0.3	0.8
3.52	4.02	4.52
2.02	2.52	3.02

8 Estimates will vary.
a 27.39 b 106.472 c 1.387 d 73.708
e 2.7 f 54.2 g 112.8 h 37.26
i 11.575 j 2.717 k 0.426 l 0.142
m 2.79 n 2.891 o 21.1 p 307

Lesson 3

1 a $-8 < -5$ b $4 > -8$ c $34 > -21$ d $28 > -10$ e $-1 < 6$
f $2 > -9$ g $-5 < 0$ h $9 > -16$ i $13 < 14$ j $-11 < -10$

2 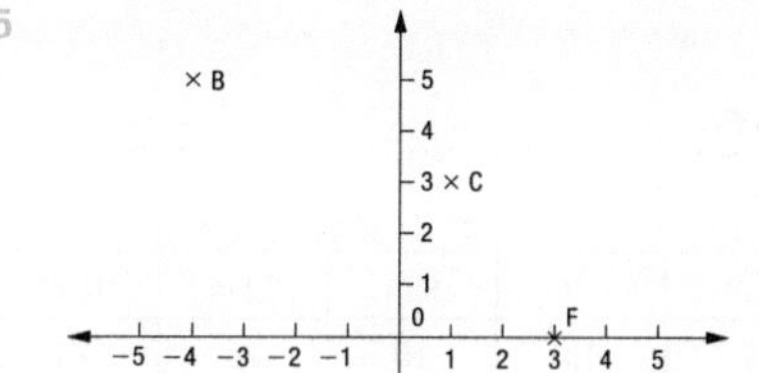

$-4 + 7 = 3$

3 a Teacher to check. b −3 c 5
d Above ground 8, 7, 6, 4, 2. Below ground −2, −4.

4 a (2, −2) b (1,1) c (3,2) d (−2, 1)
e (−4,0) f (−2, −1) g (−2, −2)

5 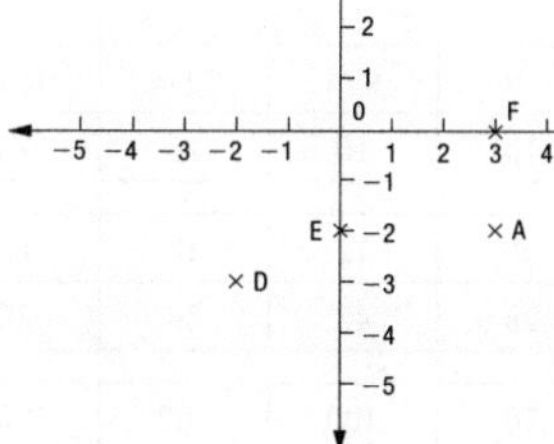

6 Teacher to check: all lines must slope upwards from left to right.

7 a 5 g per 4 kg b 40 km per litre c 2 cm per minute
d K25 per hour e 40 km per hour